T0345162

3D Printing of Sensors, Actuators, and Antennas for Low-Cost Product Manufacturing

This book discusses the 3D printing of sensors, actuators, and antennas and illustrates how manufacturers can create smart materials that can be effectively used to prepare low-cost products. The book also includes how to select the appropriate process for your manufacturing needs.

3D Printing of Sensors, Actuators, and Antennas for Low-Cost Product Manufacturing offers the most recent developments in 3D printing of sensors, actuators, and antennas for low-cost product manufacturing; the book highlights some of the commercially available low-cost 3D printing processes that have higher efficiency and accuracy. Fundamental principles and working methodologies are presented with a critical review of the past work involved and current trends with future predictions. It covers composite and polymeric materials widely used and specifically focuses on low-cost elements. Recent breakthroughs and advantages in product manufacturing when printing smart materials are also discussed.

Manufacturing engineers, product designers, manufacturing industries, as well as graduate students, and research scholars will find this book very useful for their work and studies.

Sustainable Manufacturing Technologies: Additive, Subtractive, and Hybrid

Series Editors: Chander Prakash, Sunpreet Singh, Seeram Ramakrishna, and Linda Yongling Wu

This book series offers the reader comprehensive insights of recent research breakthroughs in additive, subtractive, and hybrid technologies while emphasizing their sustainability aspects. Sustainability has become an integral part of all manufacturing enterprises to provide various techno-social pathways toward developing environmental friendly manufacturing practices. It has also been found that numerous manufacturing firms are still reluctant to upgrade their conventional practices to sophisticated sustainable approaches. Therefore this new book series is aimed to provide a globalized platform to share innovative manufacturing mythologies and technologies. The books will encourage the eminent issues of the conventional and non-conventual manufacturing technologies and cover recent innovations.

Advances in Manufacturing Technology: Computational Materials Processing and Characterization
Edited by Rupinder Singh, Sukhdeep Singh Dhami, and B. S. Pabla

Additive Manufacturing for Plastic Recycling: Efforts in Boosting A Circular Economy
Edited by Rupinder Singh and Ranvijay Kumar

Additive Manufacturing Processes in Biomedical Engineering: Advanced Fabrication Methods and Rapid Tooling Techniques
Edited by Atul Babbar, Ankit Sharma, Vivek Jain, Dheeraj Gupta

Additive Manufacturing of Polymers for Tissue Engineering: Fundamentals, Applications, and Future Advancements
Edited by Atul Babbar, Ranvijay Kumar, Vikas Dhawan, Nishant Ranjan, Ankit Sharma

Sustainable Advanced Manufacturing and Materials Processing: Methods and Technologies
Edited by Sarbjeet Kaushal, Ishbir Singh, Satnam Singh, and Ankit Gupta

3D Printing of Sensors, Actuators, and Antennas for Low-Cost Product Manufacturing
Edited by Rupinder Singh, Balwinder Singh Dhaliwal, and Shyam Sundar Pattnaik

For more information on this series, please visit: https://www.routledge.com/Sustainable-Manufacturing-Technologies-Additive-Subtractive-and-Hybrid/book-series/CRCSMTASH

3D Printing of Sensors, Actuators, and Antennas for Low-Cost Product Manufacturing

This book discusses the 3D printing of sensors, actuators, and antennas and illustrates how manufacturers can create smart materials that can be effectively used to prepare low-cost products. The book also includes how to select the appropriate process for your manufacturing needs.

3D Printing of Sensors, Actuators, and Antennas for Low-Cost Product Manufacturing offers the most recent developments in 3D printing of sensors, actuators, and antennas for low-cost product manufacturing; the book highlights some of the commercially available low-cost 3D printing processes that have higher efficiency and accuracy. Fundamental principles and working methodologies are presented with a critical review of the past work involved and current trends with future predictions. It covers composite and polymeric materials widely used and specifically focuses on low-cost elements. Recent breakthroughs and advantages in product manufacturing when printing smart materials are also discussed.

Manufacturing engineers, product designers, manufacturing industries, as well as graduate students, and research scholars will find this book very useful for their work and studies.

Sustainable Manufacturing Technologies: Additive, Subtractive, and Hybrid

Series Editors: Chander Prakash, Sunpreet Singh, Seeram Ramakrishna, and Linda Yongling Wu

This book series offers the reader comprehensive insights of recent research breakthroughs in additive, subtractive, and hybrid technologies while emphasizing their sustainability aspects. Sustainability has become an integral part of all manufacturing enterprises to provide various techno-social pathways toward developing environmental friendly manufacturing practices. It has also been found that numerous manufacturing firms are still reluctant to upgrade their conventional practices to sophisticated sustainable approaches. Therefore this new book series is aimed to provide a globalized platform to share innovative manufacturing mythologies and technologies. The books will encourage the eminent issues of the conventional and non-conventual manufacturing technologies and cover recent innovations.

Advances in Manufacturing Technology: Computational Materials Processing and Characterization
Edited by Rupinder Singh, Sukhdeep Singh Dhami, and B. S. Pabla

Additive Manufacturing for Plastic Recycling: Efforts in Boosting A Circular Economy
Edited by Rupinder Singh and Ranvijay Kumar

Additive Manufacturing Processes in Biomedical Engineering: Advanced Fabrication Methods and Rapid Tooling Techniques
Edited by Atul Babbar, Ankit Sharma, Vivek Jain, Dheeraj Gupta

Additive Manufacturing of Polymers for Tissue Engineering: Fundamentals, Applications, and Future Advancements
Edited by Atul Babbar, Ranvijay Kumar, Vikas Dhawan, Nishant Ranjan, Ankit Sharma

Sustainable Advanced Manufacturing and Materials Processing: Methods and Technologies
Edited by Sarbjeet Kaushal, Ishbir Singh, Satnam Singh, and Ankit Gupta

3D Printing of Sensors, Actuators, and Antennas for Low-Cost Product Manufacturing
Edited by Rupinder Singh, Balwinder Singh Dhaliwal, and Shyam Sundar Pattnaik

For more information on this series, please visit: https://www.routledge.com/Sustainable-Manufacturing-Technologies-Additive-Subtractive-and-Hybrid/book-series/CRCSMTASH

3D Printing of Sensors, Actuators, and Antennas for Low-Cost Product Manufacturing

Edited by
Rupinder Singh, Balwinder Singh Dhaliwal, and Shyam Sundar Pattnaik

CRC Press
Taylor & Francis Group
Boca Raton London New York

CRC Press is an imprint of the
Taylor & Francis Group, an **informa** business

First edition published 2023
by CRC Press
6000 Broken Sound Parkway NW, Suite 300, Boca Raton, FL 33487-2742

and by CRC Press
4 Park Square, Milton Park, Abingdon, Oxon, OX14 4RN

CRC Press is an imprint of Taylor & Francis Group, LLC

ISBN: 9781032046808 (hbk)
ISBN: 9781032046815 (pbk)
ISBN: 9781003194224 (ebk)

DOI: 10.1201/9781003194224

Typeset in Times
by KnowledgeWorks Global Ltd.

Contents

Preface

This book provides comprehensive insights into recent developments in 3D printing of sensors, actuators, and antennas for low-cost product manufacturing. Since 3D printing provides incomparable design and manufacturing independence, this book may be useful in low-cost product manufacturing of functional prototypes with customized properties. The authors illustrate how smart materials can be effectively used to manufacture low-cost, in-house sensors, actuators, and antennas. This book covers recent innovations in the field of smart material 3D printing and their potential applications in sensors, actuators, and antennas. There is a high volume of available scientific literature advocating for various 3D printing processes which makes it problematic to choose the best suitable process for a specific application. This book also highlights some of the commercially available low-cost 3D printing processes used to process smart materials with higher efficiency and accuracy in a comprehensive way, which will appeal best to academic researchers and commercial manufacturers. This book features wide coverage of the low-cost, in-house prepared sensor materials along with their processing technologies including fundamental principles and working methodology. Also, this gives a critical review of the past work on 3D printed sensors, actuators, and antennas as well as the current trends and future research directions. This book also covers various composite and polymeric materials, widely used in the current scenario for several engineering and other allied applications with a special focus on low-cost product manufacturing.

Despite the best of our efforts there are bound to be some mistakes. The same may kindly be brought to our notice for rectification. Any other suggestions to improve the book will be thankfully acknowledged.

We sincerely wish that the book may come up to the expectations of the readers on this vital subject of additive manufacturing technology.

Rupinder Singh
Balwinder S. Dhaliwal
S.S. Pattnaik

About the Editors

Dr. Rupinder Singh is a Professor in the Department of Mechanical Engineering, National Institute of Technical Teacher Training and Research, Chandigarh. He has received PhD in Mechanical Engineering from the Thapar Institute of Engineering and Technology, Patiala. His area of research is additive manufacturing, composite filament processing, rapid tooling, metal casting, and plastic solid waste management. He has co-authored more than 400 science citation indexed research papers, 10 books, and more than 200 book chapters, and has presented more than 100 research papers in various national/international journals. His research has been cited more than 10471 times with 'H' index of 48. As per Stanford University, he has been listed among the world's top 2% scientists.

Dr. Balwinder Singh Dhaliwal is working as an Associate Professor in the Department of Electronics and Communication Engineering, NITTTR, Chandigarh, India. He received his Doctorate in Electronics and Communication Engineering from Punjab Technical University, Jalandhar, India. His research interest includes Antenna Design and Fabrication, smart sensors, and process optimization. He has more than 20 years of experience in teaching and research. He has published/ presented more than 140 research papers in reputed International/National Journals and Research conferences.

Prof Shyam Sundar Pattnaik is presently the Director of the National Institute of Technical Teachers Training and Research (NITTTR), Chandigarh. Before joining the post of Director, Prof Pattnaik served as Vice-Chancellor of the Biju Patnaik University of Technology, a Technical University of Govt. of Odisha. Prof Shyam Sundar Pattnaik completed his Post-Doctoral Research from the University of Utah, USA, and PhD in Engineering (Electronics and Telecommunication Engineering) from Sambalpur University, Odisha. Under the able supervision and guidance of Prof S. S. Pattnaik, 23 (Twenty-three) students have completed their PhDs. He has 52 ME thesis to his credit and has published about 280 papers in journals and conferences of repute. He has two international books and four international book chapters to his credit. A recipient of BOYSCAST FELLOWSHIP from DST, Govt. of India, Prof Pattnaik also is warded with SERC Visiting Fellowship, INSA Visiting Fellowship, UGC Visiting Fellowship, Distinguished Faculty Award, National Educational Excellence Award, Dr. APJ Abdul Kalam Award for Excellence, Bharat Excellence Award, Institutional Leader Excellence Award, Certificate of Commendation for excellent efforts in sponsored research scheme, Life Time Achievements in Technical Education, Merit certificate (National scholarship) Best paper Awards, etc. He is a Life Fellow of IETE, Life Member of ISTE, and Senior Member of IEEE, USA. Prof Pattnaik is a known researcher in the field of antenna and soft computing and their application to biomedical.

Contributors

Inderpreet Singh Ahuja
Punjabi University
Patiala, India

Abhishek Barwar
National Institute of Technical Teachers
Training and Research
Chandigarh, India

Anish Das
National Institute of Technical Teachers
Training and Research
Chandigarh, India

Balwinder S. Dhaliwal
National Institute of Technical Teachers
Training and Research
Chandigarh, India

Chahat Jain
I.K. Gujral Punjab Technical
University
Jalandhar, India
and
Guru Nanak Dev Engineering
College
Ludhiana, India

Jayant G. Joshi
Government Polytechnic
Nashik, India

Mandar P. Joshi
R.H. Sapat College of Engineering,
Management Studies,
and Research
Nashik, India

Atul M. Kulkarni
National Institute of Technical
Teachers Training and Research
Chandigarh, India

Sanjeev Kumar
University Institute of Engineering
and Technology
Panjab University
Chandigarh, India

Vinay Kumar
Guru Nanak dev Engineering
College
Ludhiana, India
and
University Centre for Research and
Development
Chandigarh University
Mohali, India

Isha Malhotra
Global Institute of Technology and
Management
Gurgaon, India

Ravindra A. Pardeshi
Institute of Chemical Technology
(ICT)
Mumbai, India

Shyam S. Pattnaik
National Institute of Technical
Teacher Training
and Research
Chandigarh, India

Garima Saini
National Institute of Technical Teacher
Training and Research
Chandigarh, India

Amrinderpal Singh
University Institute of Engineering
and Technology
Panjab University
Chandigarh, India

Rupinder Singh
National Institute of Technical Teachers
    Training and Research
Chandigarh, India

Ekta Thakur
Chandigarh University
Mohali, Punjab, India

Yang Wei
Nottingham Trent University
Nottingham, U.K.

1 Introduction to 3D Printing of Sensors, Actuators, and Antennas for Low-Cost Product Manufacturing

Jayant G. Joshi[1], Mandar P. Joshi[2],
Balwinder S. Dhaliwal[3], and Shyam S. Pattnaik[3]
[1]Government Polytechnic, Nashik, India
[2]R.H. Sapat College of Engineering, Management Studies, and Research, Nashik, India
[3]National Institute of Technical Teachers Training and Research (NITTTR), Chandigarh, India

CONTENTS

1.1 INTRODUCTION

A transducer known as an antenna transforms radiofrequency (RF) signals into alternating current and back again. Every communication system's transmitter and reception parts require an antenna. Around the globe, marvelous research and development are taking place to get better communication of information and to achieve a faster data rate. Recently, the human being called the wearer (user) who wears the communication systems or end devices needs body-worn antennas to establish effective and efficient communication (Joshi et al., 2010, 2012, 2014). Under such circumstances, the antenna plays a vital role to transmit or receive microwave/RF signals. These antennas are application-specific in that it is having a resonant frequency, number of frequency bands, high gain, larger bandwidth, and high directivity in miniaturized size. Microstrip patch antennas (MPAs), Wearable/textile antennas, Metamaterial antennas, fractal antennas, etc. with their feeding techniques are widely used. Much research has been explored to design and fabricate these antennas.

DOI: 10.1201/9781003194224-1

1

3D printing is found to be a promising method to fabricate the antennas with greater manufacturing flexibilities and improved antenna performance.

This chapter provides an in-depth and extensive survey of top research papers about 3D printed antennas published by international research groups in journals, magazines, and conferences of high repute. Another name for 3D printing is additive manufacturing. This is an advanced fabrication technique used to produce complex antenna structures that provide the required gain, bandwidth, and radiation pattern and also satisfy the criterion of a smaller size of less than a quarter wavelength. A wide variety of materials, including dielectrics, metals, ceramics, polymers, and bio-compatible materials, are used in 3D printing. This technique is found to be an effective technique than conventional methods to manufacture electromagnetic or microwave structures such as antennas, filters, etc. The fabrication of electromagnetic devices from electrically conductive structures with complicated shapes and dimensions is one of the most significant uses of 3D printing. A 3D object is the end product of layering many thin films. 3D printing permits the fabrication of microwave circuits, antennas, and complicated hardware. Initially, the plastic part is created, and then it is covered with a metallic cover called metallization. The goal of this chapter is to provide information to readers about various types of 3D printed antenna structures with parameters like application-specific operating frequency along with desired parameters such as gain, bandwidth, directivity, and polarization have been presented. Examples of these antennas are; microstrip patch, dipole, monopole, Bow-tie, helical, fractal, lens, spiral, etc. These antennas are manufactured by advanced 3D printing techniques.

The microstrip patch antenna's basic design consists of (a) a dielectric substrate with two metallic surfaces: a radiating patch on the top surface, and (b) a ground plane that is fed through to the antenna using different feeding methods. The desired resonance frequency, gain, and bandwidth may influence the size and structure of the radiating patch. Under 3D printing or additive manufacturing technology, these kinds of antenna structures can be produced utilizing a variety of materials and printing procedures.

The sections listed below comprise this chapter. Section 1.1 presents the introduction and details of 3D printed antennas. In Section 1.2, an extensive study of research papers is presented. In Section 1.3, the chapter is concluded with significant remarks on fabrication techniques of 3D printed antenna. Finally, the references are listed in Section 1.4.

1.2 LITERATURE SURVEY

Belen and Mahouti (2018) present microstrip quasi-Yagi (MQY) antenna operating frequency range 670–3000 MHz is fabricated using 3D printed technology. The polylactic acid (PLA) material is used to fabricate this antenna. The antenna performance is good with a gain of 3.5–4.6 dB. The manufactured antenna has omnidirectional radiation, a compact design, broadband impedance matching, and a low price. CEL Robox micro-manufacturing platform is used for prototyping this antenna.

To create a 3D package for electronics and wireless sensor network (WSN) nodes, additive manufacturing technology is used (Kimionis et al., 2015). To build complicated 3D unique structures for HF applications, it combines additive manufacturing processes. A prototype of 3D printed packaging was manufactured using an Object Connex 2603D Polyjet Printer. The cube is made using VeroWhite and the hinges are made of Grey 60 thermoset shape memory polymers (SMP).

The 3D printed tuneable CP microstrip patch antenna is developed by Farooqui et al. (2019) using the MakerBot Replicator2 printer. The frequency tuning is performed by employing four varactors between the midpoint of the four corners of the square patch and the ground. A two-layer substrate of acrylonitrile butadiene styrene (ABS) serves as the antenna's base permittivity 2.7 and loss tangent 0.005 with a thickness of 9 mm. L-slot is used in the square patch to realize circularly polarized (CP) radiation. The tunable frequency is achieved from 2.36 to 1.64 GHz. The patch's effective length changes due to the capacitance of the varactor diode, allowing for frequency control. The minimum capacitance of the varactor diode is 0.32 pF.

The additive manufacturing or 3D manufacturing method has been used by Molaei et al. (2018) to fabricate a Pyramidal horn antenna (PHA). This 3D printed antenna is lightweight and robust. With a relative permittivity of 2.8 and a loss tangent of 0.04, the PHA is modeled using 50 μm of silver on top of a 2.4 mm thick layer of VeroWhitePlus dielectric material. Using the UG-387/U flange interface, a pyramidal horn was coupled to a rectangle waveguide, designated WR-12. The PHA design is customized to fabricate the compressive horn antenna (CHA) using a dielectric pseudorandom structure that is 3D printed and inserted into the PHA. The same dielectric material as that of PHA is used to fabricate CHA. This antenna has better performance over the range of 60–90 GHz and is suitable for high-sensing-capacity, required for sensing and imaging applications.

Axial mode C-band (4–6 GHz) high gain helical antenna is developed using 3D printing having a 5 GHz center frequency by Ghassemiparvin and Ghalichechian (2019). This helical antenna is fabricated using PolyJet (Object Prime 30) 3D printing technique. The dielectric FDM printer with PLA and ABS-PC material is used. A helical antenna's metallization is done by electroless nickel (Ni) plating, then copper (Cu) electroplating. RHCP gain at the broadside for this antenna throughout the 4.1 to 5.6 GHz frequency band is greater than 12.5 dB. Utilizing 3D printing, a stacked microstrip patch array (SMPA) antenna for use in the ISM band is developed by Belen (2018). The commonly used 3D printing substrate material for this antenna used is PLA Filament-Designer Grey RBX PLA FS 391. This compound has a loss tangent of 0.005, a dielectric constant of 2.5, and thicknesses of 3 and 4.3 mm. For gain improvement, a 2 × 2 patch array is utilized. This antenna operates over the frequency range 2.4–2.5 GHz having a moderate gain of 10–11 dB.

A linearly polarized magnetoelectric (ME) dipole device based on a waveguide is presented by Ma et al. (2021). The core body of the antenna is prepared by 3D printing using dielectric material using relative permittivity of $\varepsilon_r = 2.9$ and a loss tangent of $\tan \delta = 0.01$. The antenna is composed of an E-plane T-junction, a rectangular open-ended waveguide, two rectangular thin patches, and a dielectric cube. A high-precision 3D printer from Stratasys called the Object30 was used to build this antenna.

This antenna is a strong candidate for several X-band microwave applications, including wireless and satellite communications.

The research work of a rapidly manufactured 3D printed X-band waveguide horn antenna is presented in Tak et al. (2017). Utilizing a 3D printer with fused deposition modeling (FDM) and PLA filament, this antenna was fabricated. Each part is metalized by employing conductive spray coating after 3D printing. The presented antenna is inexpensive, lightweight, and simple to prototype.

In the research work of Lu et al. (2015), 3D printing technology is employed to make the magnetic resonance imaging (MRI) RF coil housing using Stratasys, Eden Prairie, MN, USA. To reproduce the computer simulation model of the coil, copper wires are laid on top of the carved traces. To acquire the reflection coefficient (S_{11}), an L-matching circuit and a trimmer capacitor are mounted on the coil. The method of moments (MoM), 3D printing technology, and FEKO 3D computer-aided design model are used to fabricate the antennas. For 4.7-T, a quadrature birdcage coil was created to examine functional MRI of the monkey brain.

For on-body communications at microwave and millimeter-wave frequencies, a 3D printed antenna on detachable fingernails is proposed by Njogu et al. (2020). The microstrip patch antennas have been printed on an acrylonitrile butadiene styrene (ABS) removable fingernail substrate having 0.5 mm thickness, with a dielectric constant of 2.7 and tan δ of 0.0051. The rectangular patch with a microstrip transmission line and a rectangular ground plane is designed. Layers of silver ink were applied to the curved nails using aerosol jet printing and flush curing. Through a copper plating, a second copper layer was added to 28 GHz. The antennas operate at 15 GHz and 28 GHz respectively. The antenna has a 55.38° angle of curvature, giving it the appearance of a nail. By taking into account the various dielectric constants of bone, skin, fat, and nail tissues, on-finger simulations were performed to assess the performance of human tissue on antennas. This kind of antenna design can be used in IoT solutions with 5G technology and is lightweight, inexpensive, and simple to install. It can also be used as a cosmetic addition.

Luneburg dielectric lens is fabricated by Saghlatoon et al. (2020) for enhancing the highest radiation gain of an aperture antenna at X-band between 8 and 12 GHz applications. The material used to fabricate the lens is 1.75 mm thick Polyamide 6 (Nylon 6) filament using a Markforged 3D printer at 275°C. The structure has been printed from two hemisphere-shaped parts that are taped together securely. The directivity has been increased between 14.5 and 17dB in the proposed frequency band.

For 5G applications, a high-resolution selective-ink-deposition manufacturing method on complicated 3D objects, packages, and modules is provided in Palazzi et al. (2019). In this technique, the desired patterns are embossed on a 3D printed dielectric surface and the ink is applied with an appropriate tool. The MIMO antenna system is developed for 3.4–3.8 GHz in 6 GHz 5G of band applications such as hotspot and access-point applications. The dielectric constant of 2.8 and tan δ of 0.03 is used. In this work, an embossed surface patterning technique is used to facilitate the manufacturing of 3D microwave circuits using a 3D printing tool.

3D printed special type of biodegradable polymer PLA is used by Kumar et al. (2018) to design dielectric resonator antennas (DRAs). It uses biodegradable PLA with a dielectric constant (εr) of 3.45 and a loss tangent of 0.05. Three conventional DRAs: Cylindrical DRA (CDRA), Rectangular DRA (RDRA), and Triangular DRA (TRDA) with supporting beams were designed for broadband applications. These antennas are utilized for scanning and discrimination, mobile service, radio navigation, and other C-band, X-band, and Ku-band applications because of their benefits like broader impedance bandwidth, lightweight, and low cost.

Wang et al. (2018) employed the inhomogeneous substrate and superstrate developed by 3D printing to obtain dual-band, dual-circularly polarized patch antennas. To minimize their loss of tangent, the substrate and superstrate are made from photosensitive resins and air with a square mesh-grid structure. For fabrication, the photosensitive liquid polymer is sliced with a laser using Stereo lithography appearance (SLA) technology. The substrate is 3D printed and is transparent having a dielectric constant of 3.11 and tan δ of 0.0253 is used. The photosensitive material is layer by layer cured for the fabrication process. The fabricated antenna operates at 2.75 and 3.2 GHz. It is a dual-band, dual-circularly polarized antenna that is small, lightweight, and advantageous for satellite communication systems.

The design, simulation, and fabrication of Hilbert fractal antenna using 3D printing processes are explained by Johnson et al. (2019). The structure is simulated using a 3D electromagnetic ANSYS HFSS simulator. Hilbert antenna geometry is created using copper having high conductivity and reasonable strength and rigidity. The antenna is fabricated using several 3D printing processes like Laser powder-bed fusion, Binder jetting, Vat photopolymerization, Binder jetting, polymerization, plating, binder jetting of the sand mold, and casting processes. The mechanical and electrical performance of the antenna is evaluated and found to be stable. This antenna is applicable for low-frequency applications.

Details of 3D printing techniques and different materials used to manufacture different antennas are reviewed in a detailed manner by Kaur and Saini (2018). Complicated antenna structures can be produced using 3D printing that is lightweight, quick, cost-effective, and ecologically friendly. The most widely used 3D printing techniques are- Stereolithography, fused deposition modeling, and selective laser sintering. Different 3D printed antenna configurations reviewed in this research article are- (a) 3D elastic antenna from conductive acrylonitrile butadiene styrene (ABS) (b) Bowtie antenna for 7.81 GHz with coplanar waveguide feed is manufactured using materials like ABS filaments and flexible PLA for conductive parts and dielectric parts respectively. (c) Helical antenna design at 9.4 GHz with a lens is fabricated using a novel blend of inkjet printing using silver nanoparticle ink and 3D printing of acrylonitrile butadiene styrene (ABS) plastic material. (d) 2.45 GHz CP dipole antenna using thermoplastic ABS with copper is presented. (e) A WR 10 horn antenna is manufactured using conductive ABS and conductive PLA, Amphora polymer. (f) The impedance bandwidth of microstrip patch antennas has been broadened by using 3D printed Acrylonitrile Butadiene Styrene substrates. MakerBot Replicator 2X has been utilized for the 3D printing of these substrates. By adding an air cavity to substrates, the impedance bandwidth is increased by 90%. (g) A NinjaFlex substrate is used for different bending conditions to achieve flexibility applications. A ring resonator using NinjaFlex substrate is proposed. (h) A patch antenna and unfolded Origami cube is manufactured using thermoset SMPs for 2–2.6 GHz.

It is explained that the antennas manufactured by 3D printing have several applications in Bluetooth, Wi-Fi, medical and industrial segment. These antennas are lightweight, miniaturized, and efficient.

Multiple Input Multiple Output (MIMO) antennas for 5G and millimeter (mm) wave applications have been projected by Alkaraki and Gao (2020) which are manufactured by 3D printing and these antennas deliver beams in many directions for real-time elevation up to ± 30° without the use of phase shifters. These antennas operate at 28 GHz with wide bandwidth performance. MIMO antenna structure is constructed in two parts- (a) Feeding microstrip layer by conventional PCB fabrication method and (b) radiating structure fabricated by a 3D printing method. For 3D printing, Object30 3D printer is used to print a layer of thickness of 16 μm and resolution of 100 μm using verlo white material. The 4×3 steerable MIMO consisting of six elements is used.

In this research work presented by Radha et al. (2020), a low-profile electrically compact antenna that was 3D printed is proposed which has an electric meandered dipole at the center antenna with extended arcs. The arcs mimic the current flow of the loop antenna.

Hasni et al. (2019) reported a single-step process evaluation of 3D printed patch antennas using conductive and dielectric filament. The authors have investigated the substrate characteristics such as relative permittivity via experiments. The collected data shows that obtained permittivity is comparable with standard substrates used for patch antenna fabrication. It has been reported by a research group that air gaps are created while printing the 3D objects. To resolve this issue, the authors have suggested three different layer printing patterns namely parallel, perpendicular and zigzag. The perpendicular layer printing pattern has been demonstrated in the research for printing 5.8 GHz rectangular patch antennas. The research also focuses on selecting conducting material for patch and ground plane fabrication using the conductivity of the material. The conducting material used in the present research is 3D black magic due to its high conductivity of 166.7 S/m.

SLA based 3D manufactured MIMO antenna for 5G wireless communication has been proposed by Li et al. (2017). The reported process of 3D printing uses photosensitive liquid polymer as a substrate for an antenna. A Hyaline Visijet crystal polymer material having a relative permittivity of 2.85 and a loss tangent of 0.002 at 10 GHz is used in this research. Further, customized copper plating methods have been used to coat the copper layer. In copper plating, various processes such as mechanical polishing, chemical coursing, and sensitizing have been used. Finally, an electroplating process for tinning the metal layers has been adopted to fabricate the antenna sample. As the reported research work uses polymer material for substrate fabrication, low-temperature soldering technology is implemented for the SMA connector connection.

It is revealed from the literature that, 3D printing can significantly reduce the fabrication time of large and complex antenna structures while maintaining the performance of an antenna. The antenna arrays which use a complex feeding network with a power divider are crucial in designing and fabricating. The time consumed by fabricating such an antenna array can be reduced by up to 40% as reported by Sun et al. (2020). In reported research work, an antenna array of 8 × 8 and 16 × 16 elements with the self-supporting structure for a complex feeding network has been analyzed. The 3D fabricated antenna exhibits wide impedance bandwidth of 32.4% with a peak gain of 27.1 dB. The direct metal laser sintering (DMLS) methodology has been proposed to reduce the fabrication time. This time saving has been realized by fabricating the layers without supports.

Additive Manufacturing based 3D printed single substrate multilayered stacked antenna arrays have been presented by Li et al. (2020). The research work reports detailed steps of 3D printing of multi-layered antenna structures. The reported 3D printer consists of 512 nozzles for dielectric and conductive ink jetting. Apart from this, it also uses an infrared (IR) radiation lamp and an ultraviolet (UV) lamp for sintering conductive ink and curing dielectric ink respectively. AgCite nano-silver and acrylate have been used for conductive and dielectric material respectively. The process of 3D printing has been executed at 140°C temperature. The AM approach incorporates piezoelectric-based nozzles to deposit the liquid ink for manufacturing. The fabricated multilayered antenna structure exhibits wide impedance bandwidth of 10.6% with a topmost gain of more than 3.8 dBi.

A miniaturized microstrip antenna has been presented by Patel et al. (2016) using 3D printed substrate. A planar 3D printed ridge of triangular and semi-circular shape has been fabricated on top of the substrate. This ridge helps to lower the resonance frequency of the patch and thereby realizing a miniature antenna. Commercially available PLA is used for substrate and ridge fabrication. Copper was then glued on both sides of the 3D printed substrate to realize a miniature antenna. By fabricating a ridge of the height of 1 mm, nearly 13% of the downshift in resonant frequency has been reported in research.

The popularity of fabricating antennas using a 3D inkjet printer is increasing. The academic and industrial research community is considering the 3D printing of different objects as an upcoming tool. Due to numerous advantages such as a smooth fabricated surface with layer accuracy in μm, it is becoming easy to fabricate passive microwave devices using 3D printing. A 2.4 GHz patch antenna fabricated using a 1236 inkjet nozzle-based 3D printer is reported by McKerricher et al. (2016). To increase the gain and radiation efficiency of the proposed antenna, the antenna includes a substrate in the form of a honeycomb. The fabrication process utilizes wax material, UV-cured polymer, and silver nanoparticle ink. To form a honeycomb-shaped substrate, wax and polymer are used. The reported special resolution of the 3D printer is $33 \times 33 \times 29$ μm^3. The conducting patch and the ground plane of the antenna have been fabricated using silver nanoparticles. The experimental results show that the dielectric constant varies from 3.2 to 2.8 for low MHz to 10 GHz. The 3D printed patch with a honeycomb-shaped substrate exhibits antenna radiation efficiency up to 81% with a gain of more than 8 dB at 2.4 GHz.

Flexible and wearable electronics are gaining more recognition in health care applications. Now a day, such products are becoming important in wireless bio-medical applications. Considering this recent scenario, 3D printed flexible substrate using newly introduced NinjaFlex filament material has been demonstrated for 2.4 GHz patch antenna by Moscato et al. (2016). This 3D printed flexible material shows better strain, flexibility, and printability. The substrate material of dimensions 65×55 mm^2 has been fabricated with the desired permittivity. The permittivity has been measured using the ring resonator technique with different fill characteristics. For the simulation and fabrication of a patch antenna, the relative permittivity of 3.0 with a loss tangent of 0.06 has been taken into consideration. The reported resolution of a 3D printer to fabricate flexible substrate is 50 μm. For fabrication of the reported substrate, the thickness of the planar layer of 100 μm is considered.

Farooqui and Shamim (2017) reported that a 3D inkjet printer helical antenna with a lens is integrated to enhance the gain. The printed structure comprises a combination of 2D and 3D printed geometries fabricated in a monolithic manner. The used 3D printer has a resolution of 16 μm. A photosensitive polymer named VeroBlackPlus having a relative permittivity of 2.6 and a loss tangent of 0.023 has been used for fabrication. For helix, silver nanoparticle-based ink having a conductivity of 3.8×10^5 S/m is used. The fabricated helix antenna provides a peak gain of 12.8 dBi with an end-fire radiation pattern.

An all-metallic 3D printed horn antenna is proposed by Gordon et al. (2017) for CubeSat application. The corrugated feed horn antenna has been fabricated using aluminum alloy AlSi10Mg and fabricated using the powder bed fusion (PBF) process. The reported antenna exhibits a peak gain of 38.03 dB in the broadside direction. The fabricated antenna has been tested for 118 GHz frequency.

Design compact antennas at lower UHF frequencies always remain a challenge for antenna researchers. To battle with such trade-offs, compact inverted-F antenna (IFA) using 3D printed flexible substrate has been reported by Cosker et al. (2017). The reported antenna has been fabricated using 3D printing technology. NinjaFlex material having a relative permittivity of 2.6 and a dissipation factor of 0.04 has been used to manufacture the reported antenna. The radiating element has been fabricated using Galinstan liquid metal, the substrate of an antenna is fabricated using flexible NinjaFlex material and electro-textile copper has been used for a ground plane. The reported inverted-F antenna has been designed and optimized to resonate at 885 MHz UHF frequency having a peak gain of 3.6 dB.

Bjorgaardet al. (2018) demonstrated material description using a coaxial transmission line to evaluate the relative permittivity of substrate for 3D printing. Three types of antennas such as fractal bow-tie, Yagi- Uda, and spiral-shaped have been designed, made up using a 3D-printer, and tested. The presented results show improved impedance bandwidth for fabricated antennas. In reported work, PLA material has been used for 3D fabrication.

Belen et al. (2020) reported that the implementation frequency selective surface (FSS) loaded DRA using 3D printing technology. PLA material having an infill rate of 66% with relative permittivity of 2.4 is used to fabricate the dielectric material of and above the radiating element. The reported antenna resonates in the 2.4 GHz ISM band with gain and bandwidth of 3.6 dBi and 700 MHz respectively. The 3D printing technology of the reported antenna helps to reduce the size, cost, and overall volume of an antenna with high accuracy.

Wang et al. (2021) have proposed a tolerance analysis approach based on interval arithmetic (IA) for heterogeneous 3D printing. Four important factors, including patch length, patch width, substrate height, and material qualities, are taken into account in this research. These parameters are considered as interval variables and based on errors the equations between error, resonant frequency, and E/H radiation patterns are derived. The sample antennas are fabricated using a self-developed heterogenous 3D printer and measurements are taken. The heterogeneous 3D printer consists of ink injection and laser sintering for radiating element fabrication and UV resin injection and curing are employed for substrate fabrication. The fabricated antenna uses substrate material having a relative permittivity of 2.7 with a dissipation factor of 0.016. The substrate thickness of 1 mm is considered while designing and fabricating.

The fused deposition modeling (FDM) printing technique is presented by Colella et al. (2020) to prepare substrates for antennas and microwave components using acrylonitrile butadiene styrene (ABS) or PLA materials. Metamaterials, metasurfaces, megastructures, and plasmonic metamaterials have been fabricated using 3D printing technology. In this paper, a 3D printed multibeam compact circularly polarized (CP) spherical lens antenna with 2D ultrawide angle coverage is reported. This antenna provides stable performance in X-band and is suitable for satellite communication applications. This study provides information on how to raise dielectric constants while preserving low losses to get around the limitations of typical 3D-printable materials. It presents various configurations of RFID antennas with their enhanced designs. Firstly, 3D printing technology is used to fabricate substrate with PLA to realize a T-resonator device useful for RFID designs. Secondly, Barium Titanate ($B_aT_iO_3$) is united with a Silicone matrix to obtain high permittivity flexible substrates using 3D printing models. A bracelet-shaped UHF RFID tag is fabricated using this technique. The permittivity of a substrate can be enhanced using a PLA matrix doped with $B_aT_iO_3$.

A compact triangular slot cut microstrip fed monopole microstrip antenna (MA) design is proposed by Bicer and Aydin (2021). 3D printing technology is used to fabricate printed curved substrates for biomedical applications. The commercial Creality Ender 3 Pro printer was used to produce the curved PLA substrate. Initially, the curved substrate was prepared and the copper tape was glued as a conductive element on the front and back sides of the substrate. The Sparrow Search algorithm (SpaSA) is used to optimize the antenna geometry. This antenna provides a wide bandwidth (−10dB) between frequency ranges of 3 and 12 GHz.

1.3 CONCLUSIONS

In this Chapter, different 3D printing or additive manufacturing techniques to fabricate different types and configurations of antennas have been discussed. 3D printing is a useful technique to fabricate complicated and miniaturized antennas, MIMO antennae, fractal, conformal antenna structures for Wi-Fi, Wi-Max, Bluetooth, BAN, Satellite Communication, and personal communication applications. This technique provides better flexibility and characterization to select the substrate of desired dimensions like dielectric constant (ε_r), thickness (h), and loss tangent (tan δ). Further, it provides liberty to select suitable feeding techniques for an antenna. This Chapter provides an extensive literature survey of different types of 3D printers, materials, and fabrication techniques adopted by different research groups to design and fabricate antenna and antenna array structures. 3D printing technique offers the features like creating complex structures, excellent material selection capability, flexible designs so that users can create dielectric as well as conductive/radiating patches and better mechanical support and strength. 3D printing technique is found to be suitable to miniaturize the antenna size, to enhance the gain and bandwidth along with high directivity and desired polarization of antenna elements.

REFERENCES

Alkaraki, S., & Gao, Y. (2020). mm-wave low-cost 3D-printed MIMO antennas with beam switching capabilities for 5G communication systems, IEEE Access, 8, 32531–32541.

Belen, M. A. (2018). Stacked microstrip patch antenna design for ISM band applications with 3D printing technology, Microwave Optical Technology Letters, 61, 1–4. https://doi.org/10.1002/mop.31603

Belen M. A, & Mahouti P. (2018). Design and realization of quasi-Yagi antenna for indoor application with 3D printing technology, Microwave and Optical Technology Letters, 60, 2177–2181.

Belen, M. A., Mahouti, P., & Palandoken, M. (2020). Design and realization of novel frequency selective surface loaded dielectric resonator antenna via 3D printing technology, Microwave, and Optical Technology Letters, 62, 2004–2013.

Bicer, M. B., & Aydin, E. A. (2021). A novel 3D-printed curved monopole microstrip antenna design for biomedical applications, Physical and Engineering Sciences in Medicine, 44, 1175–1186.

Bjorgaard, J., Hoyack, M., Huber, E., Mirzaee, M., Chang, Y. -H., & Noghanian, S. (2018). Design and fabrication of antennas using 3D printing, Progress in Electromagnetic Research C, 84, 119–134.

Colella, R., Chietera, F. P., Montagna, F., Greco, A. & Catarinucci, L., (2020). Customizing 3D printing for electromagnetics to design enhanced RFID antennas, IEEE Journal of Radio Frequency Identification, 4, 452–460.

Cosker, M., Lizzi, L., Ferrero, F., Staraj, R., & Ribero, J. -M. (2017). Realization of 3D flexible antennas using liquid metal and additive printing technologies, IEEE Antennas and Wireless Propagation Letters, 16, 971–974.

Farooqui, M. F., & Kishk, A. (2019). 3D-printed tunable circularly polarized microstrip patch antenna, IEEE Antennas, and Wireless Propagation Letters, 18, 1429–1432.

Farooqui, M. F., & Shamim, A. (2017). 3D inkjet-printed helical antenna with integrated lens, IEEE Antennas and Wireless Propagation Letters, 16, 800–803.

Ghassemiparvin, B., & Ghalichechian, N. (2019). Design, fabrication, and testing of a helical antenna using 3D printing technology, Microwave Optical Technology Letters, 62, 1–4. https://doi.org/10.1002/mop.32184

Gordon, J. A., Novotny, D. R., Francis, M. H., Wittmann, R. C., Butler, M. L., Curtin, A. E., Guerrieri, J. R., Periasamy, L., & Gasiewski, A. J. (2017). An all-metal 3D-printed cubesat feed horn, IEEE Antennas and Propagation Magazine, 59, 96–102.

Hasni, U., Green, R., Filippas, A. V., & Topsakal, E. (2019). One-step 3D printing process for microwave patch antenna via conductive and dielectric filaments, Microwave and Optical Technology Letters, 61, 734–740.

Johnson, K., Zemba, M., Conner, B. P., Walker, J., Burden, E., Rogers, K., Cwiok, K. R., Macdonald, E., & Cortes, P. (2019). Digital manufacturing of pathologically-complex 3D-printed antennas, IEEE Access, 7, 39378–39389.

Joshi, J. G., & Pattnaik, S. S. (2014). Metamaterial loaded microstrip patch antennas, Book Chapter; Book Title: "Encyclopedia of Information Science and Technology," Third Edition, Category: Networking and Telecommunications, Chapter 613, pp. 6219–6238 Information Science Reference (an imprint of IGI Global), USA. DOI: 10.4018/978-1-4666-5888-2, ISBN13: 9781466658882, ISBN10: 1466658886, EISBN13: 9781466658899.

Joshi, J. G., & Pattnaik, S. S., (2014). Metamaterial-based wearable microstrip patch antennas, Book Chapter; Book Title: *Handbook of Research on Progressive Trends in Wireless Communications and Networking*, Chapter 20, pp. 518–556, Information Science Reference (an imprint of IGI Global), USA. DOI: 10.4018/978-1-4666-5170-8, ISBN13: 9781466651708, ISBN10: 1466651709, EISBN13: 9781466651715.

Joshi, J. G., Pattnaik, S. S., & Devi, S. (2012). Metamaterial embedded wearable rectangular microstrip patch antenna, International Journal of Antennas and Propagation, Special issue on Wearable Antennas and Systems, 2012, 1–9, article ID 974315.

Joshi, J. G., Pattnaik, S. S., Devi, S., & Lohokare, M. R. (2010). Electrically small patch antenna loaded with metamaterial, IETE Journal of Research, 56, 373–379.

Kaur, A., & Saini, G. (2018). 3D-printed antennas: A Review, International Journal of Engineering Science and Computing, 16582–16586.

Kimionis, J., Isakov, M., Koh, B. S., Georgiadis, A., & Tentzeris, M. M. (2015). 3D-printed origami packaging with inkjet-printed antennas for RF harvesting sensors, IEEE Transactions on Microwave Theory and Techniques, 63, 4521–4532.

Kumar, P., Dwari, S., Utkarsh, Singh, S. & Kumar, J. (2018). Investigation and development of 3D-printed biodegradable PLA as compact antenna for broadband applications, IETE Journal of Research, 66, 53–64.

Li, M., Yang, Y., Iacopi, F., Nulman, J., & Ram, S.C. (2020). 3D-printed low-profile single substrate multi-metal layer antennas and array with bandwidth enhancement, IEEE Access, 8, 217370–217379.

Li, Y., Wang, C., Yuan, H., Liu, N., Zhao, H. & Li, X. (2017). A 5G MIMO antenna manufactured by 3D printing method, IEEE Antennas and Wireless Propagation, 16, 657–660.

Lu, H., Sun, X., Bolding, M. S., Reddy, C. J., & Wang, S., (2015). Fast prototyping of near-field, antennas for magnetic resonance imaging by using MOM simulations and 3D printing technology, IEEE Antenna, and Propagation Magazine, 57, 261–266.

Ma, Z. L., Chan, C. H., & Chen, B. -J., (2021). A 3D-printed waveguide-based linearly polarized magnetoelectric dipole antenna, IEEE Antenna, and Wireless Propagation Letters, 20, 68–72.

McKerricher, G., Titterington, D., & Shamim, A. (2016). A fully inkjet-printed honeycomb-inspired patch antenna, IEEE Antennas, and Wireless Propagation Letters, 15, 544–547.

Molaei, A., Bisulco, A., Tirado, L., Zhu, A., Cachay, D., Dagheyan, A. G., & Martinez-Lorenzo, J. (2018). 3D-printed e-band compressive horn antenna for high-sensing-capacity imaging applications, IEEE Antenna and Wireless Propagation Letters, 17, 1639–1642.

Moscato, S., Bahr, R., Le, T., Pasian, M., Bozzi, M., Perregrini, L., & Tentzeris, M. M. (2016). Infill-dependant 3D-printed material based on ninjaflex filament for antenna applications, IEEE Antennas, and Wireless Propagation Letters, 15, 1506–1509.

Njogu, P., Sanz-Izquierdo, B., Elibiary, A., Jun, S. Y., Chen, Z., & Bird, D. (2020). 3D-printed fingernail antennas for 5G applications, IEEE Access, 8, 228711–228719.

Palazzi, V., Su, W., Bahr, R., Bittolo-Bon, S., Alimenti, F., Mezzanote, P., Valentini, L., Tentzeris, M. M. & Roselli, L. (2019). 3D printing-based selective-ink-deposition technique enabling complex antenna and RF structure for 5G applications up to 6 GHz, IEEE Transactions on Component, Packaging and Manufacturing Technology, 9, 1434–1447.

Patel, S. S., Zuazoia I. J. G., & Whittow, W. G. (2016). Antenna with three-dimensional 3D-printed substrate, Microwave, and Optical Technology Letters, 58, 741–744.

Radha, S.M., Shin, G., Park, P., & Yoon, I. -J. (2020). Realization of electrically small, low-profile quasi-isotropic antenna using 3D printing technology, IEEE Access, 8, 27067–27073.

Saghlatoon, H., Honari, M. M., Aslanzadeh, S. & Mirzavand, R., (2020). Electrically-small Luneburg lens for antenna gain enhancement using 3D printing filing technique, International Journal of Electronics and Communications (AEU), 124, 153352.

Sun, F., Li, Y., Ge, L., & Wang, J. (2020). Millimeter-wave magneto-electric dipole antenna array with a self-supporting geometry for time-saving metallic 3D printing, IEEE Transactions on Antennas and Propagation, 68, 7822–7832.

Tak, J., Kang, D. -G., & Choi, J., (2017). A lightweight waveguide horn antenna made via 3D printing and conductive spray coating, Microwave Optical Technology Letters, 59, 727–729.

Wang, C., Li, P., Xu, W., Song, L., & Huang, J. (2021). Tolerance analysis of 3D-printed patch antennas based on interval arithmetic, Microwave, and Optical Technology Letters, 63, 516–524.

Wang, S., Zhu, L.& Wu, W. (2018). 3D-printed inhomogeneous substrate and superstrate for application in dual-band and dual-CP stacked patch antenna, IEEE Transactions on Antennas and Propagation, 66, 2236–2244.

2 3D Printing of Sensors, Actuators, and Antennas
Materials and Processes

Chahat Jain[1,2], Balwinder S. Dhaliwal[3], and Rupinder Singh[3]
[1]I.K. Gujral Punjab Technical University, Jalandhar, India
[2]Guru Nanak Dev Engineering College, Ludhiana, India
[3]National Institute of Technical Teachers Training and Research, Chandigarh, India

CONTENTS

2.1 INTRODUCTION

The last few decades have seen a tremendous rise in the application of plastic-based products due to economical reasons and further prediction indicates that there will be huge availability of non-biodegradable plastic in the coming years (Laria et al., 2020). Traditional methods of disposing of plastic shave created never-ending problems for society (Kumar et al, 2019). Also, it has been observed that the methods that are commercially available for recyclability and reusability of plastic are quite expensive. Hence, this raises an urgent requirement for low-cost, environmentally safe plastic recyclability and reusability method which prove as long-term reliable solutions to save the environment from the danger of non-recyclable plastic abundance. Latest research in this field includes the development of products such as sensors that can be used and mounted conformally for various applications at a reasonable cost (Jain et al., 2021; Ahmed et al., 2017; Malek et al.,2017). Earlier research included the design of sensor prototypes using conventional rigid materials like FR-4, glass, etc. (Njoku et al., 2011; Cook and Shamim, 2012; Abutarbosh and Shamim, 2012). Also, flexible materials like different types of cloth such as denim, nylon, and paper-based materials have been implemented for the design of antennas/sensors (Mansour et al., 2015; Ahmed et al., 2015). However, such kinds of materials had either limitations of rigidity or had major changes in the functionality in case of environmental degradation. It included exposure to humidity, heat, dust, water, etc. In contrast, the majority of polymer-based substrates don't show any change in their behavior on exposure to various environmental factors.

The majority of the research has been conducted for sensor usage in single frequency applications such as Bluetooth or any other military application. Hitherto, little has been reported on the design testing of the antenna as a sensor for a multiband utility where one single sensor design can be made to work for different applications. Such innovative designs in the polymer reusability field can serve as a base for future 3D printed IC packages as well as the development of 3D printed actuators. These can serve the required purpose in the case of soft robotics where the shape manipulations can help to achieve the predicted behavior. Research gaps have been found by using open-source software (VOS viewer). Figures 2.1–2.3 highlight the gap in the usage of primarily recycled polymers for sensor applications in BlueTooth and WiFi.

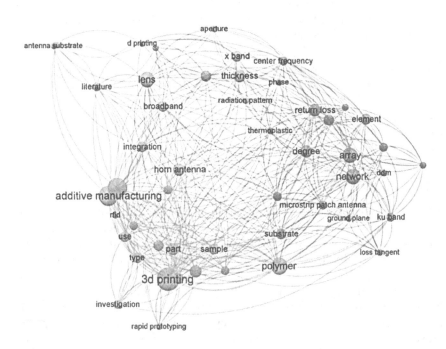

FIGURE 2.1   A gap analysis in 3D printing of antennas across the full web.

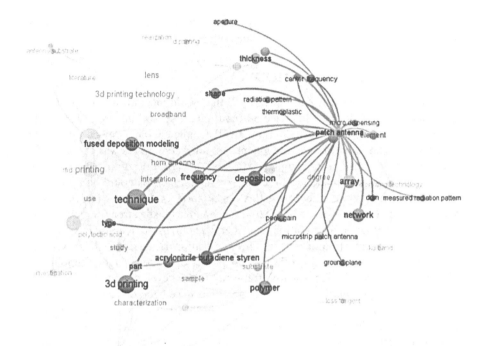

FIGURE 2.2 Research gap between the process of 3D printing and antenna designing.

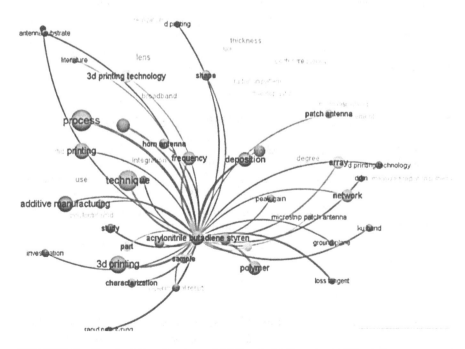

FIGURE 2.3 Gap depicting the usage of ABS material for antenna 3D printing.

In this chapter, primarily recycled polymer namely ABS has been investigated for its utility in multiband (in the range of Bluetooth and WiFi).

2.2 MATERIALS AND METHOD

To build an environment-friendly prototype that could meet all the requirements of a ready-to-fit-in device, the selected material as sensor substrate was ABS. As the main focus was to develop a practically implementable device that could solve the problem of non-biodegradable plastic utilization, ABS was chosen to be a good option because of its abundant availability. Hence designing, fabrication, and testing of devices made with the help of such materials could be done with ease, making it an economical investigation to carry on. Another reason to use ABS was its material properties. Due to its acceptable tensile strength, good resistance to environmental degradation, and chemical corrosion, it could easily serve the purpose of application of sensors in tough conformal terrains where weather and environment play a major role in their proper operation. A good amount of effort has been made by the research fraternity in the development of such antennas made out of polymeric waste. Also, antennas have been designed on flexible substrates like cloth, paper, and FR-4 type rigid substrates which are operable at different frequencies. Hitherto, little has been explored on the design and testing of polymeric waste-based antennas/sensors which could operate at more than one frequency. i.e. the same antenna could achieve multi-band functionality. This has been achieved by the introduction of truncations. In this study, single-sided conducting copper tape with a thickness of 0.08 mm has been used to develop the conducting parts of the corner truncated antenna namely, the patch and the ground plane. To provide excitation to the antenna, SubMiniature version A (SMA) (Female, Right angle) F R/A edge connector has been soldered. Figure 2.4 describes the detailed analysis as well as the design method followed for the corner truncated antenna.

2.3 ABS-BASED 3D PRINTED SENSOR: A CASE STUDY

The initial step toward the development of a 3D printed sensor is the choice of material that is to be used for additive manufacturing. In this chapter, the sensor was designed and fabricated for a frequency of 2.45 GHz with the technique of 3D printing using the locally procured ABS material. The motive behind choosing primary recycled ABS was its feature of 100% recyclability. Initially, the material was tested for its rheology and radio-frequency (RF) characteristics, and then the developed prototype was tested for its morphological and RF characteristics.

2.3.1 MELT FLOW TEST

The building up of a 3D printing process starts with the rheological property analysis of the material which is to be 3D printed. By using the ASTM D1238 standard for MFI values, the rheology of ABS polymer was analyzed.

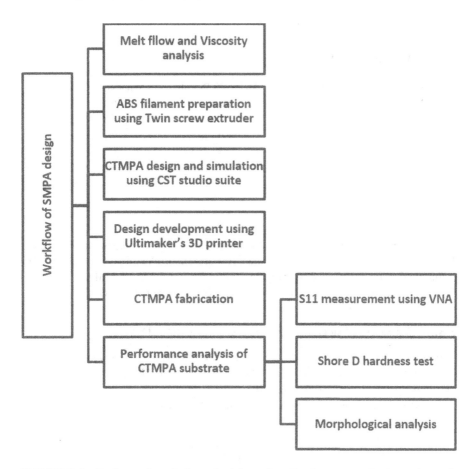

FIGURE 2.4 Design and analysis methodology flow for the truncated microstrip patch antenna.

With the standard load conditions (3.8 kg) and temperature of the die (230 °C), the material's viscosity was calculated through the recorded values of weight and volume. Equation (2.1), depicts the viscosity calculation, where the test load is denoted by 'L' and the density is denoted by 'ρ' (Sharma et al., 2020). Table 2.1 gives the details of the rheology test performed on the procured ABS polymer (Source: Batra polymers, Ludhiana, India)

$$\mu = \frac{9.13 \times 10 \times 4L \times \rho}{1.83 \times MFI} \text{ dyne/s} \tag{2.1}$$

2.3.2 TWIN SCREW EXTRUSION (TSE)

The granules of locally procured ABS polymer were extruded at 230°C using the Thermofisher's twin-screw extruder (HAAKE miniCTW). The wire-shaped

TABLE 2.1
ABS Rheological Analysis

S.No.	Melt Flow Index (g/10 min)	Material's Density (g/cm³)	Material's Viscosity (in Pa-s)
1)	10.1	1.093×10^{-3}	70932.5
2)	11.1	1.450×10^{-3}	76352.4
3)	14.0	1.466×10^{-3}	61573.4
4)	11.3	1.203×10^{-3}	42001.3
5)	10.2	1.004×10^{-3}	70545.2

specimen obtained using this method was directly provided as input to the 3D printer. The preparation of ABS filament is shown in Figure 2.5. The prepared wire spool is maintained to be in the uniform range of diameter ~2 mm. These uniform diameter ABS filaments were further utilized to provide input to the Ultimaker's 3D printer.

2.3.3 TRUNCATED MICROSTRIP PATCH ANTENNA (MPA) DESIGN

Through the parameters provided in Table 2.2, the MPA was designed and simulated in CST microwave studio suite 2019. The set of Equations (2.2)–(2.5) describes the step-by-step calculations (Balanis, 2016).

Initially, the microstrip patch width is found by using equation 2:

$$W = \frac{c}{2f_r} \sqrt{\left(\frac{2}{\epsilon_r + 1}\right)} \tag{2.2}$$

where,
W= Patch width,
Effective dielectric constant, ϵ_{eff} is evaluated using the following equation:

$$\epsilon_{eff} = \left[\frac{\epsilon_r + 1}{2}\right] + \left[\left(\frac{\epsilon_r - 1}{2}\right)\left(1 + 12\frac{h}{W}\right)^{-0.5}\right] \tag{2.3}$$

Preparation of ABS wire using TSE

Specimen of uniform ABS wire

FIGURE 2.5 (a) Extrusion process of ABS filament (b) Extruded ABS filament.

TABLE 2.2
Parametric Design Values for Corner Truncated MPA

Parameter	Value
Resonant frequency, f_r	2.45GHz (Bluetooth)
	5GHz (WiFi)
Substrate's length	65 mm
Substrate's width	65 mm
Substrate's height (h)	0.75 mm
Conductor thickness (t)	0.08 mm
$D_k,(\epsilon_r)$	2.404
Tanδ (Dissip. factor)	~0.0045
Feedline width	2.2 mm
SMA connector impedance	50 Ω

To evaluate the exact MPA's length, the fringing factor, ΔL is calculated:

$$\Delta L = 0.42h \left[\frac{\left(\epsilon_{eff} + 0.3\right)\left(\frac{w}{h} + 0.264\right)}{\left(\epsilon_{eff} - 0.258\right)\left(\frac{w}{h} + 0.8\right)} \right] \tag{2.4}$$

Thus, the length of a patch is calculated as:

$$L = \frac{1}{2 f_c \sqrt{\left(\epsilon_{eff} \epsilon_0 \mu_0\right)}} - 2\Delta L \tag{2.5}$$

Where, ϵ_0 and μ_0 represents the standard values for air's permittivity and permeability.

The final dimensions obtained are W=45 mm, and L=37 mm.

Figure 2.6 shows the simulative corner truncated microstrip patch antenna design. A corner truncation with dimensions (5 mm×5 mm) is introduced at the

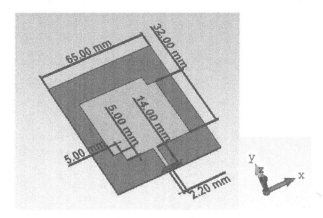

FIGURE 2.6   Corner truncated MPA in CST microwave studio.

ABS substrate
Copper tape conducting patch
Corner truncation
Microstrip feedline

Copper tape
ground plane

SMA F R/A
connector

FIGURE 2.7   Corner truncated MPA (a) front view (b) back view.

top right and bottom left of the MPA for achieving resonance in more than one band. Impedance matching is obtained by using a microstrip line feed excitation with width=2.2 mm.

Square-shaped truncations were inscribed in the top right and bottom left corners of the patch to multiple resonances. The length (L_{tr}) and width (W_{tr}) of the truncation are varied from 1 mm to 6 mm using the feature of parametric sweep in CST.

2.3.4 ANTENNA FABRICATION

Through fused deposition modeling, the corner truncated MPA substrate was developed. For this particular design, the infill pattern used was rectilinear, infill density was set to be 100% and the raster angle of 45° was chosen to print the antenna substrate of 0.75 mm height. Figure 2.7 shows a 3D printed prototype of corner truncated MPA.

2.3.5 ANTENNA TESTING

The dimensions, L_{tr}=5 mm (length) and W_{tr}=5 mm (width) for the top right and bottom left corner truncations respectively resulted in good return loss values as shown in Figure 2.8.

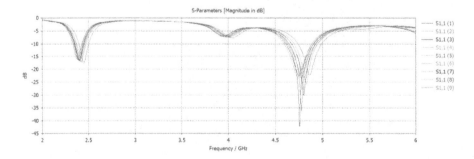

FIGURE 2.8   Parametric sweep for length and width of corner truncations.

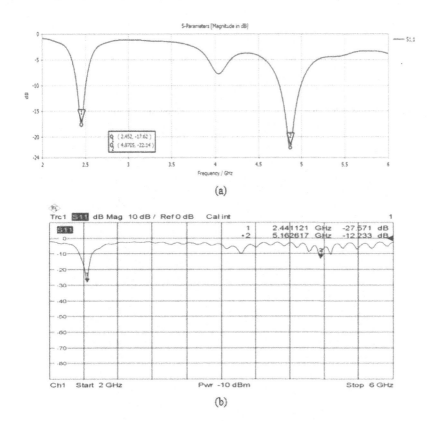

FIGURE 2.9   (a) Simulated S_{11} value (b) Experimental S_{11} value.

Figure 2.9 depicts the simulated and experimental results. After the calibration of VNA for open circuit, short circuit, and load matching, VNA is deployed to verify the return loss (S_{11}) and voltage standing wave ratio of the fabricated antenna. The analysis of S-parameters and VSWR was performed for the frequency range 2 to 6 GHz which covers both, Bluetooth as well as Wi-Fi operability.

2.4.   RESULTS AND DISCUSSION

2.4.1   MORPHOLOGICAL AND MECHANICAL CHARACTERIZATION

Table 2.3 presents the detailed material characterization (morphological as well as mechanical) of the additively manufactured substrate for corner truncated microstrip patch antenna designed for utilities in Bluetooth and WiFi-operated devices. The porosity level recorded for the surface finish was 22.55% and for the cross-sectional view, it was found to be 25.68%. This guaranteed that a very less number of air voids have incurred during the 3D printing process. This further ascertained quite a

TABLE 2.3
Detailed Morphological and Mechanical Characterization

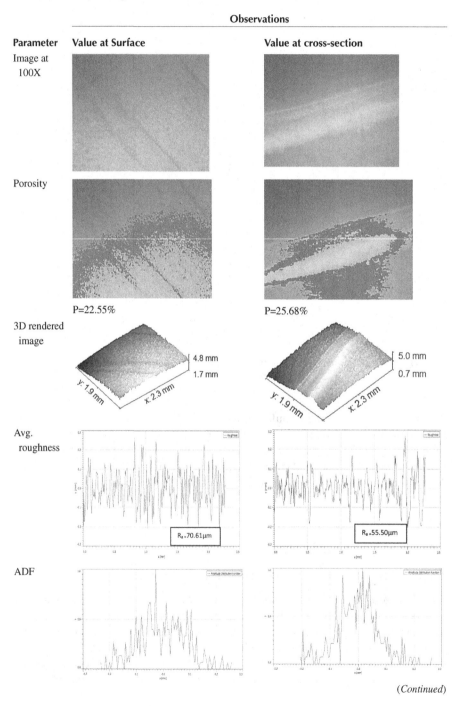

	Observations	
Parameter	**Value at Surface**	**Value at cross-section**
Image at 100X		
Porosity	P=22.55%	P=25.68%
3D rendered image	4.8 mm / 1.7 mm / y: 1.9 mm / x: 2.3 mm	5.0 mm / 0.7 mm / y: 1.9 mm / x: 2.3 mm
Avg. roughness	R_a=70.61μm	R_a=55.50μm
ADF		

(Continued)

TABLE 2.3 (*Continued*)
Detailed Morphological and Mechanical Characterization

	Observations	
Parameter	Value at Surface	Value at cross-section
BRC	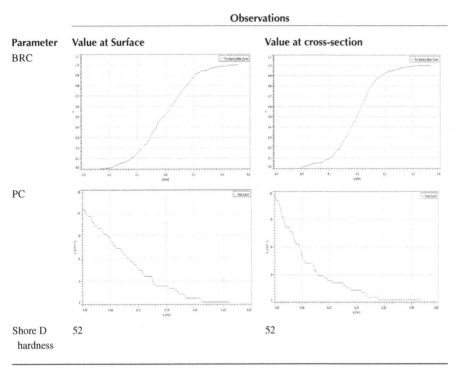	
PC		
Shore D hardness	52	52

low value of dissipation factor (0.0045 approx). A low value of average surface roughness (70.61 μm) was recorded, thereby supporting a good antenna gain and good radiation efficiency. The surface hardness of the non-biodegradable ABS polymer-based substrate was on averagely recorded to be 52 using the Shore-D Durometer.

2.4.2 RF CHARACTERIZATION

The 3D printed corner truncated antenna fabricated with the use of non-biodegradable ABS polymer was also analyzed for various conformal applications. Figures 2.10–2.15 show the comparison of return loss between simulated and fabricated structures at different angles of bending. It is seen that the antenna performs within the required range of operability up to a certain angle of bending, thus proving its utility to human wearable applications. The radiation analysis has been further supported by the current distribution pattern at 2.4 GHz as per Figure 2.16 with the maximum value of current 41A approximately. The 3D radiation plot of the antenna shows a maximum directivity of 7.033 dBi as per Figure 2.17.

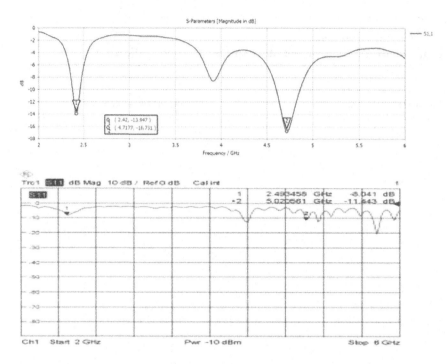

FIGURE 2.10 S_{11} for antenna bent at 10° concave position (a) Simulated (b) Experimental.

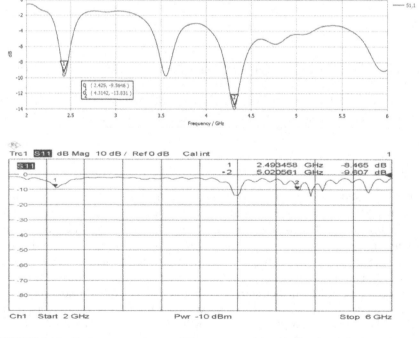

FIGURE 2.11 S_{11} for antenna bent at 10° concave position (a) Simulated (b) Experimental.

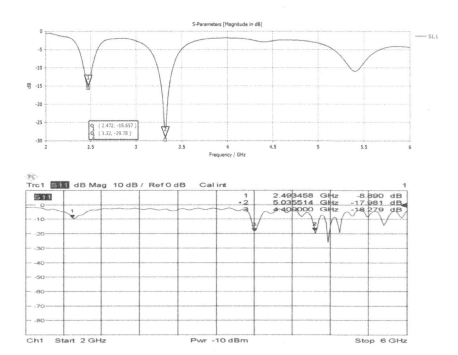

FIGURE 2.12 S_{11} for antenna bent at 30° concave position (a) Simulated (b) Experimental.

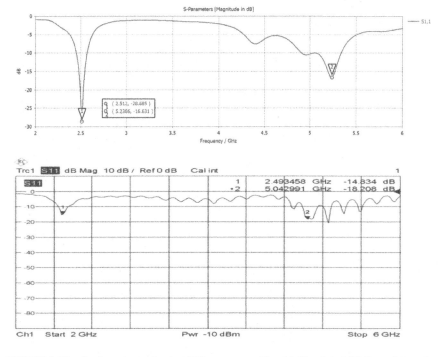

FIGURE 2.13 S_{11} for antenna bent at 10° convex position (a) Simulated (b) Experimental.

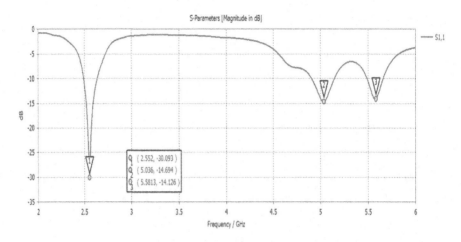

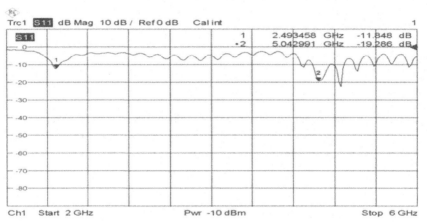

FIGURE 2.14 S_{11} for antenna bent at 20° convex position (a) Simulated (b) Experimental.

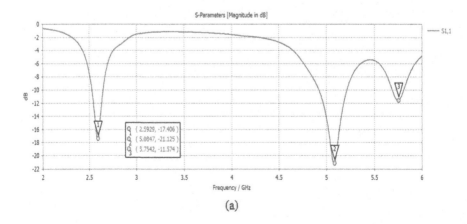

(a)

FIGURE 2.15 S_{11} for antenna bent at 30° convex position (a) Simulated.

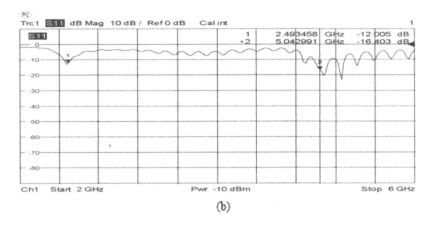

(b)

FIGURE 2.15 (*Continued*) S_{11} for antenna bent at 30° convex position (b) Experimental.

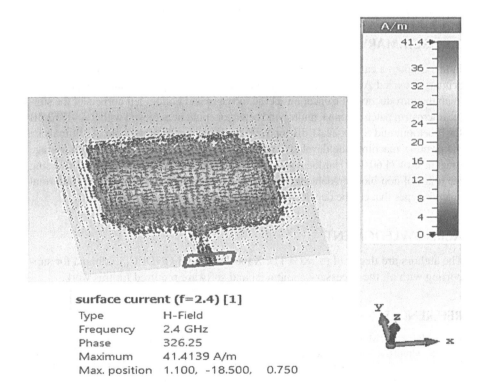

surface current (f=2.4) [1]

Type	H-Field
Frequency	2.4 GHz
Phase	326.25
Maximum	41.4139 A/m
Max. position	1.100, -18.500, 0.750

FIGURE 2.16 Simulated surface current distribution at 2.4 GHz.

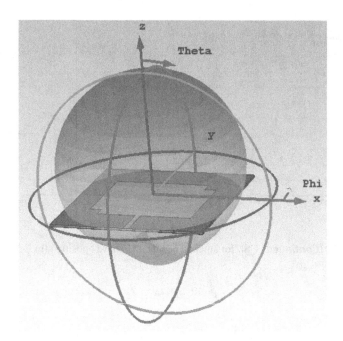

FIGURE 2.17   3D radiation pattern at 2.4 GHz (directivity=7.033 dBi).

2.5 SUMMARY

In this chapter, a case study of the design and parametric investigation of a 3D printed primary recycled ABS-based corner truncated microstrip patch antenna is presented. With the introduction of truncations at the top right and bottom left corners of the simple microstrip patch antenna, multi-band behavior could be achieved with S_{11}= -17.62 dB for Bluetooth and S_{11}= - 22.41 dB for WiFi applications. With the help of Ultimaker's 3D printing machine, the developed substrate had a porosity of 22.66% and average roughness of 74.60 µm. Hardness was averagely found to be 52 (Shore D), thus stressing the usage of non-biodegradable thermoplastics such as ABS for flexible and conformal antenna types that can be deployed in different flexible packages.

ACKNOWLEDGMENT

The authors are thankful to IKGPTU, Kapurthala, and GNDEC, Ludhiana for supporting with all the necessary equipment and software required for this work.

REFERENCES

Abdul Malek, N., Mohd Ramly, A., Sidek, A., & Yasmin Mohamad, S. "Characterization of acrylonitrile butadiene styrene for 3D-printed patch antenna," Indones. J. Electr. Eng. Comput. Sci., vol. 6, no. 1, pp. 116–123, 2017, doi: 10.11591/ijeecs.v6.i1.pp116-123.

Abutarboush, H. F., & Shamim, A. (2012). Paper-based inkjet-printed tri-band U-slot monopole antenna for wireless applications. IEEE Antennas and Wireless Propagation Letters, 11, 1234–1237.

Ahmed, M. I., Ahmed, M. F., and Shaalan, A. A. "Investigation and Comparison of 2.4 GHz Wearable Antennas on Three Textile Substrates and Its Performance Characteristics," pp. 110–120, 2017, DOI: 10.4236/ojapr.2017.53009.

Ahmed, S., Tahir, F. A., Shamim, A., & Cheema, H. M. (2015). A compact kapton-based inkjet-printed multiband antenna for flexible wireless devices. IEEE Antennas and Wireless Propagation Letters, 14, 1802–1805.

Balanis, Constantine A. *ANTENNA THEORY ANALYSIS, AND DESIGN*, 4th edition. John Wiley & Sons, Inc., Hoboken, New Jersey, 2016.

Cook, B. S., & Shamim, A. (2012). Inkjet printing of novel wideband and high gain antennas on low-cost paper substrate. IEEE Transactions on Antennas and Propagation, 60(9), 4148–4156.

Jain, C., Dhaliwal, B. S. & Singh, R. (2021). Flexible and Wearable Patch Antennas Using Additive Manufacturing: A Framework. Reference Module in Materials Science and Materials Engineering, 10.1016/B978-0-12-820352-1.00093-6.

Kumar, S., Singh, R., & Batish, A. "On investigation of rheological, mechanical and morphological characteristics of waste polymer-based feedstock filament for 3D printing applications," 2019, DOI: 10.1177/0892705719856063.

Laria, G., Gaggino, R., Peisino, L. E., & Cappelletti A. "Mechanical and processing properties of recycled PET and LDPE-HDPE composite materials for building components," 2020, DOI: 10.1177/0892705720939141.

Mansour, A. M., Shehata, N., Hamza, B. M., & Rizk, M. R. M. (2015). Efficient design of flexible and low-cost paper-based inkjet-printed antenna. International Journal of Antennas and Propagation, 2015.

Njoku, C. C., Whittow, W. G., & Vardaxoglou, J. C. (2011). Effective permittivity of heterogeneous substrates with cubes in a 3D lattice. IEEE Antennas and Wireless Propagation Letters, 10, 1480–1483.

Sharma, R., Singh, R., & Batish, A. (2020). On effect of chemical-assisted mechanical blending of barium titanate and graphene in PVDF for 3D printing applications. Journal of Thermoplastic Composite Materials, DOI: 10.1177/0892705720945377.

3 3D Printing in Antenna Design

Atul M. Kulkarni[1], Garima Saini[1],
Shyam S. Pattnaik[1], and Ravindra A. Pardeshi[2]

[1]National Institute of Technical Teacher
Training and Research, Chandigarh, India
[2]Institute of Chemical Technology (ICT), Mumbai, India

CONTENTS

DOI: 10.1201/9781003194224-3

3.1 INTRODUCTION

The antenna is integral to a wireless communication system (Balanis, 1992). In the last decade, AM was enormously used to develop RF and microwave circuits. Manufacturing antennas with different characteristics and almost any geometry is made possible by 3DP. Advent in 3DP technology provided enormous advantages compared to conventional antenna manufacturing techniques, which include faster product development cycle, the capability of more complex and flexible design, i.e., unconventional shape antenna, reduction in prototyping time and cost, less human interaction, etc. A more compact antenna design as compared to a conventional 2D planner patch antenna is made possible because of 3DP, which enables the integration of chip-antennas (Sravani & Rao, 2015).

Tuning of effective permittivity of dielectric substrate or material is possible with the help of structuring a dielectric material with a 3D lattice design, i.e., this technology made it possible to adjust the effective dielectric constant of dielectric material to satisfy the needs of antenna design (Mazingue et al., 2020). Another simplest way to change the material's refractive index in AM process is to change the printing parameter like infill percentage (Poyanco et al., 2022). Graphene is also proven a superior choice for the conductive part of antennas because of the features like biocompatibility, thermal conductivity, zero bandgaps, and carrier mobility with high surface area (Akbari et al., 2014; Mirzaee & Noghanian, 2017).

This chapter explores various 3D printed antennas designed for different applications. Various AM techniques used in 3D printed antennas are presented here. The recent antenna structures like corrugated pyramidal horn, dielectric reflectarray, 3D compact on-chip antenna, horn array, corrugated horn, tunable CP microstrip patch antenna, switchable 5-element bow-tie antenna array, axial corrugated, mm-Wave dielectric rod antenna, conical horn, etc. produced with the AM method has been discussed. A class of various thermoplastic polar polymers like ABS, PLA, and PETG utilized by various researchers, along with other probable potential polar polymers which are not used in the literature related to 3D printed antennas, are explored in this chapter.

Revolutionary antenna realization concepts made possible by 3DP technology along with recent and future surface coating materials and metallization techniques used for antenna design are presented afterward. Opportunities, future scope, current challenges, and possible, probable solutions in 3D printed antenna technology concerning nanomaterials, nanocomposites, and SMP are discussed subsequently.

3.2 3D PRINTING TECHNOLOGIES: OVERVIEW

In 1986, stereolithography (SLA) first 3DP technology, was patented by Charles Hull. Then afterward, in the year 1988, selective laser sintering, i.e., SLS by Carl Deckard. Later in the year 1992, Scott Crump created the first functional Fused Deposition Modeling (FDM) printer (Bensoussan, 2016). Commercially available basic 3DP AM techniques are extrusion, selective sintering and melting, powder binder bonding, polymerization, and layer laminate manufacturing (LLM) (Andreas Gebhardt, 2012).

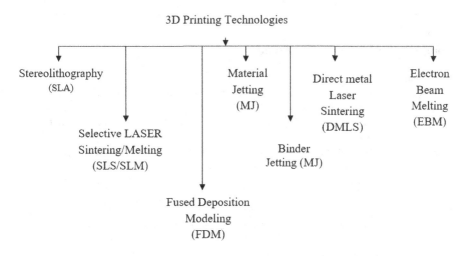

FIGURE 3.1   3D Printing technologies.

Evolution in several 3DP technologies took place concerning the material state or type, functionality, appearance of the end product, or mechanical properties of the end product. Some of these 3DP technologies related to antenna applications are shown in Figure 3.1. The generalized step of AM to generate individual physical layers and combine them to obtain the end 3D product is stated as follows,

- 3D model creation of an antenna or substrate using simulation software
- Generating standard triangulation language (STL) format of the model
- 3D Slicing (generate G code), i.e., breaking down the design into several layers
- Building layer by layer to create a design using a 3D printer
- Post-processing, i.e., cleaning and finishing the model

3.2.1   AM Techniques Summary

With so many AM technologies, the most used 3DP technologies are SLA, SLS, and FDM. The comparison of these technologies is given in Table 3.1.

3.3   3D PRINTING FOR ANTENNAS

Earlier realization and characterization of any arbitrarily complex shape antenna was not possible, but the advancement of 3DP technologies helped researchers to realize almost any geometry with low material wastage. Traditional antenna parameters like gain can be enhanced by integrating 3D printed parts (Lens et al., 2021). 3DP technologies can design all-metal, dielectric, or microstrip patch kinds of antennas. In the subsequent sections, the use of various AM techniques for the realization of 3DP antennas is reviewed concerning the different 3DP antenna types. Various kinds of the antenna operating at different frequency bands with different materials

TABLE 3.1

Comparison of Basic AM Techniques

Technology	FDM	SALE	SLS
Material Feed	Plastic Filament Example: Standard thermoplastics such as ABS, PLA	Polymer (Light-sensitive) resin Example: Varieties of resin (thermosetting plastic)	Metal and Polymer powder Example: Nylon stainless steel, ABS
Process	Thermoplastic Polymer filament melting and extrusion	Laser curing and solidifying the photopolymer resin	Powder of polymer, metal, or resin is sintered by using a Laser
Merits	Fast Low-cost consumer machine	High accuracy Smooth surface finish	No need for the support structure
Demerits	Low accuracy	High cost Not Echo Friendly	Rough surface finish Limited material options

printed by AM technologies such as SLA, fused deposition modeling, conductive ink printing, Electron beam melting, material jetting, and powder binder bonding are explained.

3.3.1 ELECTROMAGNETIC MATERIALS FOR 3D PRINTING ANTENNA APPLICATIONS

At present, very few 3D printable conductive polymers, thermoplastic polymers, or conductive materials such as ABS, Acrylonitrile styrene acrylate (ASA), PLA, PETG, Polydimethylsiloxane (PDMS), Thermoplastic polyurethane (TPU), nylon, resin, polymethyl methacrylate (PMMA),VeroWhite, Black Magic 3D, Electrifi, Stainless steel 316 L, Aluminum, Cu-15Sn, etc. are used in 3DP antenna applications. These limited materials (dielectric or metal) used for 3DP restricted the number of applications related to electromagnetic properties.

A solution methodology of mixing nanoparticles with a 3D printable polymer matrix is presented in (Liang et al., 2014). This methodology further can be used to artificially control effective material properties like permittivity, permeability, and loss tangent. Some 3D printed antennas require metallization. Many researchers explored metallization methods and various metal coating materials (Hoel, Kristoffersen, et al., 2016; Lomakin et al., 2018; Mirmozafari et al., 2018; Ramade et al., 2012). The typical material options explored in the 3DP antenna design are listed in Table 3.2.

3.3.2 CLASSIFICATION OF 3D PRINTED ANTENNAS

3DP antennas can be categorized into three types depending on the materials used in manufacturing. 1. Dielectric antenna 2. Metal coated 3D printed dielectric antenna, i.e., metalized dielectric printed antennas, and 3. Metal printed antennas. Figure 3.2 depicts the classification of 3D printed antennas.

TABLE 3.2

Typical Materials (Conductive or Dielectric) Used in 3D Printed Antennas

Material	Type	Properties	Reference
Cu–15Sn	Conductive	Surface roughness= 2.79 μm	(Zhang et al., 2016)
Nylon 680	Dielectric	εr=2.67 @10 GHz tan δ= ~0.01	(Mirzaee & Noghanian, 2017)
VeroBlackPlus	Dielectric	εr=2.6 @ 9 GHz tan δ= 0.023	(Farooqui & Shamim, 2017)
Protopasta	conductive	Resistivity= 15 Ω-cm	(Hasni et al., 2019)
3D Black Magic	conductive	Resistivity= 0.6 Ω-cm	(Hasni et al., 2019)
ABS	Dielectric	εr=2.7 @ 1.67–2.17 GHz tan δ= 0.005 εr=2.56 @ 1.2–1.4 GHz tan δ= 0.008	(Farooqui & Kishk, 2019; Fenn et al., 2016)
Stainless Steel 316L	conductive	Surface roughness= 16 μm, 12.8916 μm	(Gu et al., 2020; Zhang et al., 2016)
PDMS	Dielectric	εr=2.76 @ 3–7 GHz tan δ= 0.06	(Mohamadzade et al., 2021)
PLA	Dielectric	εr=5 @2.4–3.8 GHz, tan δ= 0.015, εr=3.45 @ 7.5 GHz–31.3 GHz, tan δ= 0.03, εr=3 @ 1 GHz, tan δ= 0.02, εr=2.5 @ 1 GHz, tan δ= 0.025	(Ahmadloo & Mousavi, 2013; Chietera et al., 2022; Kumar et al., 2021; Zhu et al., 2021)

3.4 DIELECTRIC ANTENNAS

The antennas manufactured with only dielectric material are explained here. Arbitrary complex shape dielectric antennas, which can be fabricated by 3DP, are discussed in this section.

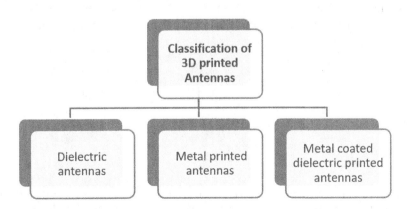

FIGURE 3.2 Classification of 3D printed antennas.

3.4.1 DIELECTRIC HELICAL ANTENNA

3D printed helical antennas based on conductor structure are reported in the literature. (S. Wang et al., 2020), proposed 3DP helical antenna with the advantage of replacing metal material with zirconia ceramic. The presented antenna does not affect the radiation performance and avoids conductor loss caused by metal material. Figure 3.3 shows three antennas with the same helical structure but having different material configurations, i.e., all-metal, dielectric with metal, and fully dielectric.

Figure 3.3 (a) represents the all-metal structure of a conventional helical antenna where the helical structure, as well as the ground, is made up of metal. Figure 3.3 (b) represents a partially metal structure, where the helical structure is dielectric, and the ground is metal. Figure 3.3 (c) represents 3D printed all-dielectric antenna, helical structure, and ground are made up of dielectric. The paper evaluated and reported the performance of all these three structures. This antenna makes full advantage of 3DP under ceramic SLA apparatus (CSLA). Actually, for such kind of antenna, the radiation increases concerning the helical shape parameters, i.e., diameter, turns, etc. The limitation of this structure is the complexity of the fabrication of turns in the conventional craft.

The reported literature evaluated the performance of three antennas in terms of antenna gain, and it was observed that the gain of the antenna with dielectric helical structure and metal ground outperform all-metal or all-dielectric antenna. There is a scope to add one more structure of this kind with the help of a dual extruder FDM 3D printer machine. The performance of a dielectric printed helical structure with the ground as a conductive polymer or carbon nanotubes (CNT)/grapheme, conductive polymer nanocomposite can be evaluated as reported in (Gnanasekaran et al., 2017).

3.4.2 DUAL-BAND ALL-DIELECTRIC REFLECTARRAY

New kind of all-dielectric dual-band reflectarray using low-cost 3DP is explained in Figure 3.4. Earlier conventional reflectarray uses metallic resonant cells or metallic ground with a dielectric slab. In contrast, this design uses air as a wideband phasing element and layered dielectric structure (transparent/mirrored) as ground. Figure 3.4 shows the basic arrangement of the Dual-band all-dielectric reflectarray. The topmost is the upper-frequency band reflectarray (3D printed using PLA) which reflects and makes a parallel beam of the V-band electromagnetic waves and allows the

(a)   (b)   (c)

FIGURE 3.3 (a) Conventional all metal antenna (Helix and Ground), (b) Partially metal structure (Dielectric Helical structure and ground as metal), and (c) All-dielectric helical structure (ground and helix) (S. Wang et al., 2020).

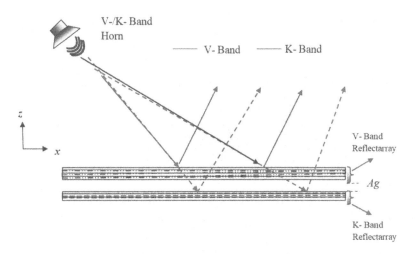

FIGURE 3.4   Basic arrangement of dual-band all-dielectric reflectarray (Zhu et al., 2021).

K-band to pass. The bottom side reflectarray (3D printed using PREPERM 3D ABS) reflects and collimates the K-band waves.

The photograph of 3D printed fabricated V-band reflectarray, K-band reflectarray, and assembled all-dielectric reflectarray is available (Zhu et al., 2021).

The basic working principle of reflectarray is based on the dielectric mirror, where a stack of dielectric layers is formed with air in between, and the propagated waves cause to reflect depending on the change in refractive index at boundaries. The dielectric layers of thickness of 0.8 mm, for the V-band, are 3D printed using PLA with a permittivity of 2.5, and loss tangent of 0.025. Similarly, K-band dielectric layers are 3D printed using ABS with a thickness of 1.5 mm, permittivity 4, and loss tangent 0.004. The reported literature evaluated the performance in terms of transmission/reflection coefficients of the V-band and K-band reflectarray concerning various layers of the dielectric structure. It is observed that reflection is directly related to the number of layers, but adding more layers increases the thickness of the antenna, hence in this design, four and five layers are adopted for K-band and V-band, respectively (Zhu et al., 2021).

3.4.3   Dielectric Resonator Antennas (DRAs)

Basically, DRAs are popular fully dielectric antennae as a radiating source, and a common method to reduce size is to introduce metal or increase the permittivity of dielectric (Rodriguez et al., 2016). One added advantage of DRAs is that it lacks metal parts. Metal becomes lossy at high frequencies. So these DRAs are more efficient than metal antennas at mm-Wave and higher microwave frequencies (Kumar et al., 2021).

Three mostly explored geometries of DRAs are rectangular, hemispherical, and cylindrical. DRA is constructed using high permittivity dielectric material mounted on the grounded low dielectric constant substrate. Generally, the dielectric resonator is fed by a microstrip transmission line. Advancements in 3DP technologies helped

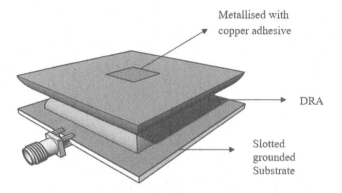

FIGURE 3.5   Rendered view of DRA (Chietera et al., 2022).

to investigate more complex types of DRAs. A wideband (2.4–3.8 GHz) DRA with an unconventional design that covers WiFi/Bluetooth as well as 5G-NR band is simulated and realized by 3DP technology, and measured (Chietera et al., 2022).

The mentioned DRA is realized by filament fused fabrication (FFF) technology using PLA to achieve a low-profile DRA design, where DRA is formed by stacking two rectangular elements having a larger size on top of other. DRA is mounted on top of the microwave feed slot. The substrate used for the microwave feed slot is commercially available PREPERM PPE1200. Interior edges of the resonator elements are rounded to increase the gain and bandwidth shown in Figure 3.5. The metallic top load on the resonator was adopted to enhance the bandwidth; the rendered view of DRA is shown in Figure 3.5.

3.4.4   DIELECTRIC PLANAR LENS ANTENNA (DPLA)

DPLA is constructed using concentric rings having different relative permittivities with equal thickness. The higher relative permittivity ring is at the center of the lens and lower at the external ring as shown in Figure 3.6 (a). After defining the center

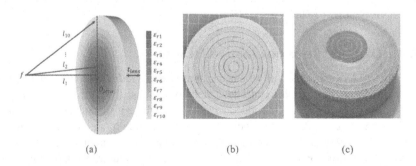

FIGURE 3.6   (a) Model of planar lens antenna having multiple concentric circles with varying permittivity, (b) Fabricated 3DP planar lens antenna, and (c) Lens antenna with matching layer (Poyanco et al., 2022).

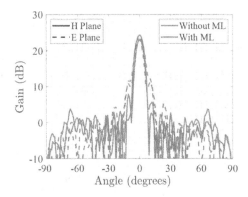

FIGURE 3.7 Performance evaluation of measured gain radiation pattern of the 3D printed dielectric lens with and without matching layer at 28 GHz (Poyanco et al., 2022).

frequency as 34 GHz, remaining parameters such as the number of rings, thickness, diameter, and the relative permittivity of each ring were calculated (Poyanco et al., 2022). The lens is constructed using the 3DP AM process, using PREPERM dielectric filaments, and the required relative permittivity is obtained by varying the infill percentage. The prepared dielectric lens antenna is shown in Figure 3.6 (b) and Figure 3.6 (c) shows the lens antenna with a matching layer. Because of the highest permittivity at the center, the reflection will be more, and the matching layers are utilized to suppress this effect. The effect of adding a matching layer to the lens can be observed by enhanced gain characteristics, as shown in Figure 3.7.

Figure 3.7 shows enhancement in measured gain due to the addition of a matching layer. however, it is also reported in the literature that, the measured gain of an antenna with a matching layer is less than the simulated result because of the constraints of the 3D printer like surface roughness, resolution porosity, etc.

3.4.5 COMPARATIVE SURVEY TABLE (DIELECTRIC ANTENNAS)

Some of the contributors who have designed and fabricated dielectric antennas with the help of 3DP are mentioned in Table 3.3.

3.5 METAL PRINTED ANTENNA

The advantage of a 3D printed all-metal antenna is that it can scale the metal structure to other frequency bands easily and hence, strongly applicable for fast prototyping and fabrication, with reduced expenses. The antennas of this kind where whole antenna parts are prepared using 3DP are explained in this section. Printing resolution and surface roughness are some challenges of the 3DP technique that needs to be tackled for future high-end applications.

3.5.1 D-BAND RESONANT CAVITY ALL-METAL ANTENNA

Mostly, horn and lens antennas are focused on mm-Wave and sub-mm Wave antennas by the 3D printer research fraternity. These antennas have large electrical sizes.

TABLE 3.3

Comparative Survey of 3D Printed Dielectric Antennas

[Ref.]	Antenna Type	Manufacturing Process	Frequency Band	Contribution
(Rodriguez et al., 2016)	Dielectric Prism Antenna (DPA)	(FDM) Makerbot Replicator 2x Printer	2.4 GHz	All-dielectric ultra-thin antenna is designed.
(Lugo et al., 2017)	Dielectric Rod Antenna	FDM (multilayer one material for core and another for cladding)	30–40 GHz Ka-band	The effect of using two thermoplastic materials one for core and another for cladding is to improve the gain by 3–8.5 dB
(S. Wang et al., 2020)	Dielectric helical antenna	CSLA	4.5 -9 GHz	All metal antenna is replaced by a dielectric with reduced cost and fast manufacturing.
(Zhu et al., 2021)	Dielectric dual-band reflectarray	FDM	K band and V band	Conventional metal-based reflectarray replaced by this design
(Kumar et al., 2021)	UWB DRA with 3D printed horn	FDM	7.5 GHz to 31.3 GHz). C to Ka-band	3DP portable multi-wavelength DRA for UWB with high gain
(Chietera et al., 2022)	DRA	Fused Filament Fabrication (FFF)	ISM Band I 2.45 GHz and 5G NR band	Compact 3D printed wideband DRA without the need fora complex metallization procedure
(Poyanco et al., 2022)	DPLA	FDMOcular3D printer	Ka-Band	Ka-band high gain, wideband dielectric planar antenna with 3D printed technology.

3D printed compact, lightweight all-metal D band antenna is presented by (Gu et al., 2020). The all-metal D Band resonant cavity-based antenna is constructed by the DMLS technique using Stainless Steel 316L. The fabricated 3D printed antenna is further polished like conventionally produced stainless steel parts because Stainless Steel 316L has excellent corrosion resistance.

The all-metal D Band resonant cavity-based antenna shown in Figure 3.8 is based on a resonant cavity antenna. The resonant cavity is created between the ground plane and superstrate. The ground plane is generally a fully reflecting structure, and a superstrate is a partially reflective surface (PRS). In this design, a 3D printed metallic grid is used as a PRS superstrate, and the impedance matching layer is used to bring the reflection coefficient below -10 dB for the desired frequency. The structure with waveguide feed is shown in Figure 3.8 (Gu et al., 2020).

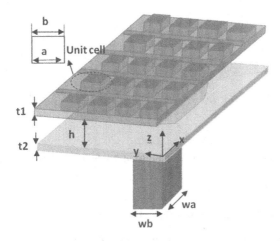

FIGURE 3.8 Resonant cavity antenna schematic view fabricated by (Gu et al., 2020).

The all-metal antenna performance is evaluated by comparing metalized dielectric printed antenna, where the dielectric printed antenna is copper plated after platinum physical vapor deposition (PVD). PVD is a vacuum metallization process where metal vapor is deposited on a dielectric structure to form a thin conductive layer. The measured result of reflection coefficient and gain concerning designed frequency for 3D printed metal and metal-coated dielectric antennas are reported where reflection coefficient is reached at −25 dB for desired frequency band as well comparative analysis plot of gain for the metal sample and polymer sample i.e., metal-coated is available (Gu et al., 2020).

The reduced gain of the metal-coated dielectric antenna in comparison to the all-metal antenna is observed because of metal loss, and it's likely due to poor metallization on the inner surface of PRS and the metal cavity.

3.5.2 DUAL-POLARIZED METAL TRANSMITARRAY ANTENNA

Transmitarray (TA) combines the feature of lens and array antenna, which helps to increase directivity and gain. A quadruple rigid waveguide (QRW) is a square waveguide with four identical ridges exactly located in the middle symmetrically shown in Figure 3.9 (b). The all-metallic 3D printed TA with high gain is constructed using QRW. The prototype schematic is shown in Figure 3.9 (a). Both TA and horn are printed using aluminum material. The 3D printed horn antenna is used as feed to the fabricated TA.

The gain of the 3D fabricated horn with and without the transmitting array is measured to evaluate the performance of the 3DP transmit array antenna. The analyzed result is available and reported by the author, which outlines that a 3D printed transmit array enhances the gain of the horn antenna by 14 dB at 30 GHz. This complex metal structure is made realizable by advancements in 3DP technology (X. Wang et al., 2022).

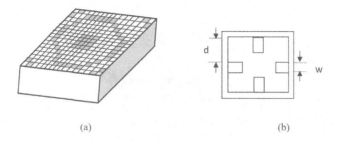

(a)   (b)

FIGURE 3.9   (a) The prototype of the fabricated TA by (X. Wang et al., 2022). (b) QRW element.

3.5.3 METALLIC 3D PRINTED CONICAL HORN ANTENNA

As compared with the conventional injection molding and micromachining, the metallic 3D printed conical horn antenna outperforms in terms of cost, production time, and environmental friendliness, and if the same is compared with non-metallic antennas manufactured by 3DP technologies these are simple and robust.

The three metal conical horn antennas with E band, D band, and H band printed by selective laser melting (SLM) technology are reported by the author, where the change in scale of the fabricated antenna concerning the design frequency can be observed. The aforementioned antennas are constructed using Cu-15Sn, which provides better surface roughness than stainless steel 316L. Two 3DP technologies, i.e., binder jetting, sintering and SLM are evaluated for stainless steel 316L and Cu-15Sn material, respectively, in terms of surface roughness. Then according to the analysis of surface roughness, Cu-15Sn is utilized for the fabrication of the aforementioned antennas (Zhang et al., 2016).

The photograph of three fabricated conical horn antennas with SLM Cu–15Sn of different sizes related to E, D, and H bands is reported by the author. The author also outlined the evaluation of antenna loss of fabricated horn antennas w.r.t. frequency. It is reported that there is good agreement between the simulated and measured results for S11 parameters of 3D printed E and D band antenna. Degradation in the response of antenna loss concerning frequency is the cause of surface roughness, as outlined by the author, i.e., the roughness of the surface causes the overall loss of SLM printed antennas (Zhang et al., 2016).

3.5.4 COMPARATIVE SURVEY TABLE (METAL PRINTED ANTENNAS)

Some of the contributors who have designed all-metal antennas with the help of 3DP are mentioned in Table 3.4.

3.6 METAL-COATED DIELECTRIC ANTENNAS

The various 3D printed dielectric antenna requires metallization. A vast number of metallization processes are available in the literature. Some of which are explored by many researchers include vacuum metallization, thermal evaporation, conductive

TABLE 3.4

Comparative Survey of 3D Printed All Metal Antennas

[Ref.]	Antenna Type	Manufacturing Process	Frequency Band	Contribution
(B. Zhang, P. Linnér, C. Karnfelt, P. L. Tarn, 2015)	Conical and Pyramidal horn	SLS (316L stainless steel)	V Band	mm-Wave 3D SLS printed conical and pyramidal horn fabricated and compared with conventional injection molded manufactured antenna.
(Rojas-Nastrucci et al., 2017)	Pyramidal horn	Binder Jetting	Ka Band	Binder jetting 3DP technology was utilized to fabricate a Ka-band pyramidal horn antenna.
(G. L. Huang et al., 2016)	Pyramidal horn	DMLS (aluminum AlSi10Mg)	X Band	Low weight 3D metal printed horn antenna is proposed. The metal sheet perforation technique is utilized to reduce the weight of the horn.
(Zhang et al., 2016)	Conical horn	SLM 316L and Cu-15Sn	E, D, and H Band	Investigation of metallic 3D printed antenna for mm-Wave and sub mm-Wave is carried out, and performance compared with nonmetallic antennas.
(Shamvedi et al., 2017)	Dual-band Sierpinski gasket	DMLS (Titanium alloy)	Multi-band (4.2 GHz and 8.4 GHz)	Sierpinski gasket complex geometry using 3DP technology (DMLS printer) proposed for easy and fast prototyping.
(Bjorgaard et al., 2018)	Fractal bow-tie	FDM, Conductive PLA	S (2.45 GHz)	With the help of different AM techniques three-antenna, the structure was designed.
(Colella et al., 2019)	Flexible tag RFID Antennas	FDM (Electrify) PLA substrate	Ultra-High Frequency (UHF)	Electrify is used with 3DP technology to quickly produce an antenna on a PLA substrate.
(Hasni et al., 2019)	Patch	FDM Black-Magic 3D, PLA	C Band	A 3D printed patch antenna using commercially available Black-Magic 3D material was proposed.
(Kotze & Gilmore, 2019)	Pyramidal horn	SLM (aluminum AlSi10Mg)	X Band	Performance of conventionally fabricated and 3D printed X band operated aluminum horn antenna evaluated.
(Reinhardt et al., 2019)	Corrugated pyramidal horn	SLM (tin bronze powder)	300 GHz	300 GHz 3D SLM printed metal (tin bronze) Corrugated pyramidal horn antenna proposed.

paint spray, arc, and flame spraying, plating (electro- and electroless), galvanic plating, etc. (Lomakin et al., 2018; Ramade et al., 2012) (Hoel, Kristoffersen, et al., 2016). The antennas of this kind where the dielectric part is prepared by utilizing 3DP are discussed here.

3.6.1 Voronoi Inverted Feed Discone Antenna and Fractal Sierpinski Triangle Antenna

3D mathematically inspired antenna structures such as complex structured Voronoi inverted feed discone antenna and fractal Sierpinski triangle antennas are proposed by (Bahr et al., 2017). The mentioned antennas are SLA printed using clear resin photopolymer and metalized using the electroless copper plating method proposed by (Fang et al., 2012). These realized antennas, which are difficult to fabricate with the conventional technique, are reported.

The partition of a plane into well-structured self-similar polygons forms the Voronoi diagram, also referred to as the Voronoi partition or Voronoi tessellation is shown in Figure 3.10 (b). This Voronoi diagram with inverted discone as the base shape shown in Figure 3.10 (a), is fabricated, and the photograph of this fabricated 3D structure is reported by the author. The 3D printed and metalized Sierpinski triangle fractal is also fabricated by the author, whose 2D prototype is shown in Figure 3.10 (c). The basic shape to reproduce self-similar structures for this antenna is an equilateral triangular pyramid. The evaluation of results outlined by the author validates that 3D fabricated both Voronoi antenna and fractal Sierpinski triangle antenna are doing well over a broad range of frequencies. The frequency shift is observed because of the variations in the fabricated feeding structures, SubMiniature version A (SMA) connector length, and its effect on the antenna structure. The author also outlined that this is the first attempt to demonstrate such a 3D mathematically inspired fractal antenna structure, which was made possible because of advancements in 3DP technology (Bahr et al., 2017).

3.6.2 Meander-Line Dipole Antenna

Integration of two steps of a metalized dielectric antenna is presented here. 3DP of dielectric and metallization is carried out by (Ahmadloo & Mousavi, 2013).

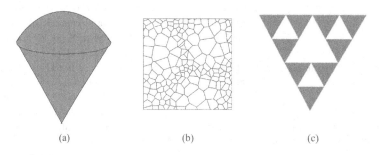

(a)   (b)   (c)

FIGURE 3.10 (a) Inverted discone structure and (b) Voronoi structure (c) Sierpinski triangle fractal 2D structure.

Intelligent Healthcare Systems

Edited by

Vania V. Estrela

Federal Fluminense University
Telecommunications Department
Rio de Janeiro, Brazil

CRC Press
Taylor & Francis Group
Boca Raton London New York

CRC Press is an imprint of the
Taylor & Francis Group, an **informa** business

A SCIENCE PUBLISHERS BOOK

First edition published 2023
by CRC Press
6000 Broken Sound Parkway NW, Suite 300, Boca Raton, FL 33487-2742

and by CRC Press
4 Park Square, Milton Park, Abingdon, Oxon, OX14 4RN

CRC Press is an imprint of Taylor & Francis Group, LLC

Library of Congress Cataloging-in-Publication Data (applied for)

ISBN: 978-1-032-05272-4 (hbk)
ISBN: 978-1-032-05274-8 (pbk)
ISBN: 978-1-003-19682-2 (ebk)

DOI: 10.1201/9781003196822

Typeset in Times New Roman
by Innovative Processors

Acknowledgments

We want to express our gratitude to several individuals who assisted us in completing this book. To begin with, we would want to express our gratitude to God. We discovered how precious this talent for writing is while putting this book together. He has endowed us with the ability to believe in our aspirations and pursue them. We could never have accomplished this without our faith in the Almighty. We do want to offer our heartfelt appreciation to all writers and co-authors for their outstanding efforts. It was commendable that we did not have to remind contributors about their submissions on a regular basis. We would like to express our gratitude to the reviewers for consenting to evaluate chapters and for their significant contributions to the chapter's quality and content presentation.

We thank the CRC team for their unwavering support. They granted us extensive deadline extensions whenever necessary. We would like to express our gratitude to everyone who has assisted us, directly or indirectly, in completing this book.

Dr. Vania V. Estrela

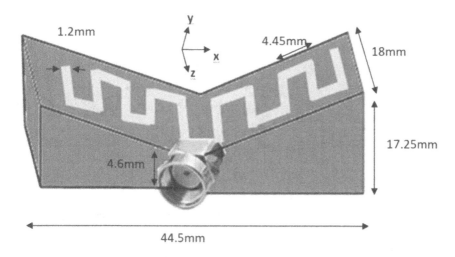

FIGURE 3.11 Prototype of fabricated meander-line dipole antenna by (Ahmadloo & Mousavi, 2013).

A new approach of simultaneously 3DP conductive ink and dielectric is discussed here. The process is explored for a meander-line dipole antenna which is printed using silver nano-particle ink on a plastic substrate having V-shaped shown in Figure 3.11. The reported work is carried out on a modified open-source FDM printer such that it is used to print ink along with the traditional polymer. The high surface porosity is achieved while printing polymer substrate to enhance structural integrity, and to print the ink on the surface of the substrate properly. The curing process of printed conductive silver ink is carried out at 85° C for 15 minutes. Well, conformity between simulated and measured S11 parameters of the designed antenna with minor change is reported by (Ahmadloo & Mousavi, 2013).

It is also outlined by the author that, the deformation of a 3D printed plastic substrate during the curing process of conductive ink is the result of a minor change in measured and simulated results. There is vast scope to advance and adapt the process for the substrate and conductive ink formulations. With the available technology, quality and accuracy of the proposed fabrication technique are evaluated and recommended, its potential for fast microwave device prototyping.

3.6.3 CORRUGATED CONICAL HORN ANTENNA

The conical horn and pyramidal horn are simple to manufacture but do not provide circularly symmetric radiation patterns. Corrugated horn is complex to manufacture by conventional method but is advantageous in terms of low side lobe levels, low cross-polarization, and circularly symmetric radiation pattern.

The complex corrugated conical horn structure having many-dimensional parameters is rapidly fabricated with low cost and low waste by SLA 3DP; the corrugated conical horn prototype structure with dimensional parameters is shown in Figure 3.12. 3D printed dielectric part of the antenna without metallization, then

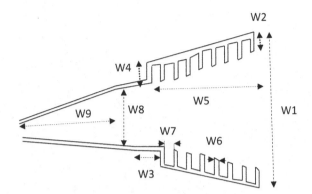

FIGURE 3.12 Corrugated conical horn model structure with dimensional parameters (Chieh et al., 2014).

after application of conductive paint, and a metal-coated assembled complete antenna reported by the author. Such type of corrugated conical horn antenna is best suitable for direct-broadcast satellite (DBS), or radio astronomy kind of applications where the power received does not exceed the maximum power handling of the thermoplastic (Chieh et al., 2014).

The antenna is printed using ABS material, and afterward, metallization is achieved using nickel aerosol paint. The author outlined the negative aspect such as uncontrolled surface roughness and non-uniformity because of manual paint spray and suggested electroplating for better results. The simulated and measured S11 (Reflection coefficient) parameters of the fabricated antenna are presented by the author. The author also noted that the spike observed in the measured reflection coefficient is probably due to the excitation of another mode that happened from the transition of rectangular to circular waveguide (Chieh et al., 2014).

3.6.4 NONPLANAR LINEAR-DIPOLE-PHASED ARRAY ANTENNAS

Six-element linear dipole array was fabricated using SLA in a single run, without the need for the assembly process and additional soldering loss compared to conventional well-established PCB technology. The conventional mentioned linear dipole array needs to mount over a separate reflector plate as a stable structure for the antenna. The author reported that this structure is made possible by 3DP with a single piece only. Reflection coefficient measurement of the one element of the array is reported by (Mirmozafari et al., 2018).

The arrangement of dipole arrays is shown in Figure 3.13 (a); the distance between each λ/2 dipole to the ground plane is λ/4. Each dipole array is fed by an SMA connector such that the inner connector of each SMA connector is proximity coupled to one pole of a dipole array element, and the outer connector is connected to the ground plane. The gap between the dipole poles is optimized and selected at 2 mm for the designed antenna. The author reports simulated and measured reflection coefficient for one dipole array element.

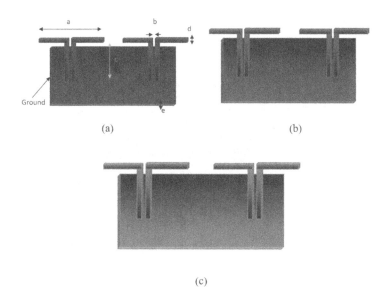

(a)

(b)

(c)

FIGURE 3.13 Fabrication process steps of array antenna prototype (a) 3D printed dielectric dipole array (b) Array after electroless plating (c) Copper-plated array antenna after electroplating (Mirmozafari et al., 2018).

Figure 3.13 shows the fabrication steps carried out to fabricate the structure of two dipole samples, Figure 3.13 (a) shows a 3D dielectric printed structure, then the same is electroless copper plated as shown in Figure 3.13 (b) and Figure 3.13 (c) shows antenna after final copper electroplating.

The result reported the feasibility of a 3D printed linear array antenna. Improved performance can be obtained using optimized antenna dimensions and an appropriate metallization technique. One added advantage of the 3DP non-planer linear dipole phased array antenna is that a continuous dielectric slab over the ground plane is not utilized hence reduction in surface waves is achieved (Mirmozafari et al., 2018).

3.6.5 PYRAMID HORN ANTENNAS

Pyramid horn is a type of horn antenna with E and H planes that are widened and used with rectangular waveguides. K band pyramid horn antenna fabricated with two stepped AM, i.e., deposition of plastic and then conductive material. The performance is evaluated by comparing a 3DP pyramid horn antenna with a conventional commercially available cast aluminum reference antenna (Lomakin et al., 2018). The Pyramid horn antenna structure is shown in Figure 3.14.

The fabricated pyramidal horn used the FDM technique to create a PLA horn structure. Then three different metallization approaches were utilized to evaluate the performance of EMC-spray deposition, galvanic electroplating, and 3D-Deposition of conductive material using a piezo-jet system. The metal material utilized for

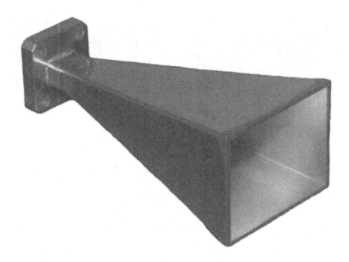

FIGURE 3.14   Structure of pyramid horn antenna.

EMC-spray, galvanic, and piezo-jet is copper, nickel, and silver, respectively. The author evaluated the aforementioned metal coating techniques in terms of weight, the width of the final product, and surface roughness, along with antenna parameters like reflection coefficient, transmission coefficient, beam pattern, half-power beam-width, etc. The 3D printed pyramidal horn has comparable performance to the full metal reference antenna with the added advantage of low weight and rapid prototyping.

3.6.6   HELICAL ANTENNA WITH INTEGRATED LENS

Helical antenna with integrated lens is manufactured by a combination of 3D printed dielectric lens and helix with 2D inkjet printing of silver nanoparticle; this digitally controlled monolithically manufactured lens integrated helical antenna dimensional parameter is shown in Figure 3.15.

VeroBlackPlus photosensitive polymer dielectric material is used to 3D print the helix and lens. To enhance the gain of conventional helix antenna, we can increase the turns but additional extra turns have a small effect on gain, so integration of lens with helix is presented here for enhancement of gain. The author reported the evaluation of simulated and measured S11 parameters and gain with and without lens at 9 GHz design frequency (Farooqui & Shamim, 2017).

The improvement in the reflection coefficient and gain for helix antenna because of Fresnel lens was observed. The author also outlined that the measured results of the S11 parameters and gain match the simulated one. Integration of helix with lens utilizing two 3D inkjet printers, one for dielectric printing and one for metal printing reflects the advancement in 3DP and its utility in producing complex antenna structures with low waste and less time compared with traditional manufacturing techniques.

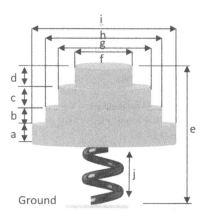

FIGURE 3.15   Lens integrated helical antenna configuration (Farooqui & Shamim, 2017).

3.6.7   COMPARATIVE SURVEY TABLE (METAL-COATED DIELECTRIC ANTENNAS)

Some of the contributors who have designed metal-coated dielectric antennas with the help of 3DP are mentioned in Table 3.5.

TABLE 3.5
Comparative Survey of 3D Printed Metal-Coated Dielectric Antennas

[Ref.]	Antenna Type	Manufacturing Process	Frequency Band	Contribution
(Chieh et al., 2014)	Corrugated conical horn	FDM, ABS material metalized with conductive ink	Ku Band (10–16 GHz)	New method to 3D print low cost and low weight Ku band corrugated conical horn antenna.
(Fenn et al., 2016)	Conformal array	FDM, ABS with copper plating	L Band	Conformal array with dielectric 3D printed and then metalized with copper plating technique proposed here.
(Bahr et al., 2017)	Voronoi and Sierpinski fractal	SLA and electroless deposition of copper	C Band (5.5 GHz)	Complex geometry 3D printed which is difficult to realize by a conventional method, and novel surface metallization method proposed.
(Toy et al., 2017)	Pyramidal horn	FDM, PLA dielectric structure later metalized process	X Band	3D FDM-based horn was fabricated using PLA material and later metalized, and this antenna was proposed for military applications.

(*Continued*)

TABLE 3.5 (*Continued*)
Comparative Survey of 3D Printed Metal-Coated Dielectric Antennas

[Ref.]	Antenna Type	Manufacturing Process	Frequency Band	Contribution
(Farooqui & Shamim, 2017)	Helical antenna with integrated lens	VeroBlackPlus dielectric and silver nanoparticle-based ink for metallization	X Band (9 GHz)	Inkjet printing is utilized to 3D print dielectric as well as metallization. The gain of the proposed antenna is increased because of lens integration with the helix structure.
(Mirmozafari et al., 2018)	6-element linear array	SLA and copper electroplating process	S-Band (2.7–3 GHz)	The proposed antenna array does not require a continuous dielectric substrate on a ground plane like patch antennas hence surface waves are greatly reduced. 3DP SLA is utilized to fabricate dielectric structure and then copper is used for metallization.
(Lomakin et al., 2018)	Pyramidal horn	FDM, PLA for dielectric structure, and later metallization	K Band (14–24 GHz)	PLA-based structure fabricated by using FDM and then EMC spray, Galvanic and piezo jet metallization method utilized and their performance compared.
(Farooqui & Kishk, 2019)	Microstrip patch (Tunable CP)	FDM (ABS) adhesive aluminum tape	L Band(1.65–2.17 GHz)	3DP enabled technique to introduce L-slot in square patch for fabrication of CP tunable microstrip patch antenna is proposed.
(Adeyeye et al., 2019)	Yagi-Uda Loop	SLA	UHF and ISM Band (2.45 GHz)	2.45 GHz yagi-uda loop antenna with the help of 3DP SLA fabricated with low cost.
(Helena et al., 2020)	Pyramidal horn	FDM (PLA) and copper metallization	Ka-Band (28 GHz)	Two metallization approaches were utilized to metal coat the 3D printed PLA-based dielectric horn structure.

3.6.8 ROLE OF NANOPARTICLES FOR 3DP ANTENNA

Nanoparticles like CNT, graphene, Montmorillonite (MMT), clay, etc. are materials with one dimension between 1 and 100 nm. The introduction of nanoparticles to 3D printed entities like microwave component fabrication is important, where enhancement of some properties such as magnetic, electrical, and optical properties is a necessity. Nanoparticles are influencing its properties like surface-to-volume ratio, lower melting point (reducing processing temperature), higher chemical reactivity (suitable in catalyst), and higher surface conductivity as compared with their counterparts in its bulk form (Goesmann & Feldmann, 2010).

When a material is in its bulk form then basically it has definite features, like color, chemical reactivity, and melting temperature. For example, the melting temperature of Gold is 1064°C, it has a yellow color, and it is not used as a catalyst when it is in its bulk form, whereas gold nanoparticles appear red to purple with a melting temperature about approx 300°C–400°C, it can be used to effectively catalyze specific reactions (Hales, 2020). That means conductive nanomaterial can be sintered at lower temperatures and this property with required modification in case of 3DP applications can be utilized, and such properties can be utilized for multi-material (metal and dielectric simultaneously printing) 3D printer.

Electrospinning is one example of a bottom-up approach, using this approach, one can generate nanoscale fibers. So in an ideal electrospinning process, when we apply a high voltage to the polymer solution and during migration from the middle tip to the collector, the solvent which is present in the solution gets evaporated, and fibers get collected on the ground collector. By varying the process parameters like voltage, flow rate, type of collector, the distance between nozzle tip to the collector, and polymeric solution, the properties of the nanofiber can be controlled and utilized for electronic applications like manufacturing of sensors, antenna, etc. The most commonly used polymer in electrospinning is an organic polymer (Y. Huang et al., 2013; Xue et al., 2019).

Carbonized cellulose nanofibers (CCNFs) are formed using the electrospinning method, and as already discussed, extrusion-based 3DP technology referred to as FFF used for creating dielectric structures of thermoplastic material, and the main issue with this process is reduced crystallinity of the end product, so to tackle this issue microwave heating is utilized. Compounded CCNFs with PLA are used to produce filament for 3DP, and further crystallinity of PLA composite can be modified by using microwave heating. This is one of the solutions to overcome crystallinity issues related to polymer (Dong et al., 2020). The synthesis of nanoparticles via top-down and bottom-up approaches is shown in Figure 3.16.

The generalized way to attain nanoparticles such as graphene thin films includes a bottom-up approach like chemical vapor deposition. From an electromagnetic material point of view, graphene possesses zero band gap with frequency-dependent conductivity this help to allow propagation of various plasmonic modes at THz frequencies, hence graphene antenna can be used to operate at Terahertz frequency band by utilizing the feature, negative imaginary conductivity of graphene (Lee et al., 2018). Multiband plasmonic antenna for THz band using graphene is reported recently (Sairam & Singh, 2022).

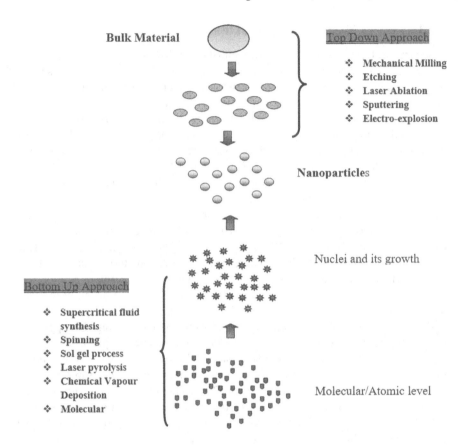

FIGURE 3.16 The synthesis of nanoparticles via top-down and bottom-up approaches (Khanna et al., 2019).

The addition of suitable nanoparticles like CNT, graphene, ZnO, nano TiO2, and MMT clay in appropriate proportion in polymer matrix leads to the formation of polymer nanocomposite. It has a good utility in the 3DP of electronics devices.

The physical phenomena involved in the AM process result in nanoparticle alignment in the printing direction. Depending on the properties of nanocomposites, thermal or electrical conductivity could be beneficial to achieve at the desired location of antenna structure while printing. The addition of nanoparticles with polymer to form polymer nanocomposite changes the rheological properties of the polymer material, this can be utilized in AM for the successful attainment of required properties of the printed structure. However, the addition of nanoparticles with the polymer may restrict the polymer chains mobility; which may cause to affect the printing quality for particular polymer composite in AM process, hence proper characterization needs to be conducted of various polymer nanocomposites to achieve desired printing with desired properties (Al Rashid et al., 2021).

There is huge scope to investigate, advanced nanomaterials like silver nanowires (AgNWs), copper nanowires (CW), Indium Tin Oxide (ITO), and Zinc Oxide (ZnO) for transparent antenna application. Highly conductive nanoparticle coating techniques, and nanocomposite polymer printing technology, for antenna fabrication, needs to be explored further.

3.7 SHAPE MEMORY POLYMERS FOR ANTENNA DESIGN

Under the influence of one of the stimuli, SMPs can come back to their original shape or change to a new shape. Typical stimuli are a magnetic field, light, temperature, the pH value of a surrounding liquid, ohmic power, and so on. https://onlinelibrary. wiley.com/doi/full/10.1002/app.50847 - app50847-bib-0009. The chemical nature of the polymer and its diverse molecular mechanisms can influence the shape memory properties of corresponding smart polymers. Along with PLA and ABS blends, other FDM printed SMPs are polycyclooctene (PCO), bisphenol, polyurethane (PU), etc. (Ehrmann & Ehrmann, 2021).

The author reported the design and manufacturing methodology of active composite which takes multiple shapes concerning change in environmental temperature, this is accomplished by 3DP with multiple families of SMP fibers, and digital SMPs with different glass transition temperatures (Tg) to influence the transformation of the structure (Wu et al., 2016).

SMPs are prepared by various techniques. Recently 3DP process based on a Vat Photopolymerization of a SMP was investigated to produce customized smart and complex morphable antennas. The highest shape recovery performance is noted by a combination of 20% of an elastomeric resin in a thermoset UV system. 3D printed helical SMP antenna with copper electroplating is fabricated and its RF performance examined, and the result shows that a thermal stimulus is capable of obtaining a conformal shape of the antenna, which can be utilized further for multi-band morphing system or adaptive antenna system (Cersoli et al., 2022).

There is a need to investigate and explore various SMPs to design and 3D print reconfigurable antennae which can lead to smart antenna systems. SMP-based all-dielectric antennas will play important role in future applications, whereas, one major challenge with metal-coated SMP antennas is the durability of metallization.

3.8 CHALLENGES, OPPORTUNITIES, AND FUTURE SCOPE

One of the major challenges in terms of 3D printable thermoplastic material is the reduced crystallinity because of a sudden change in temperature of the material when material exits from the hot nozzle and solidifies quickly at the low-temperature platform, which leads to weak mechanical performance. The probable efficient to enhance the crystallinity of 3DP polymer material.

As we are moving toward utilizing mm-Wave and THz frequency band, and FDM 3DP materials like ABS, PLA, Nylon, Bendlay are highly absorbing in the THz range whereas polystyrene acts as the best transparent in the THz band (Busch et al., 2014), these individual properties can be explored further in antenna design in mm-Wave and THz frequency band to suit the required application.

An optically transparent structure can be fabricated using the SLA process to manufacture transparent antennas or by coating transparent nanomaterial like Indium Tin Oxide (ITO) and Zinc Oxide (ZnO) on FDM printed transparent dielectric substrate.

Postprocessing is commonly involved in treating surface roughness in 3D printed parts of the antenna such as micromachined process (MMP), gold electroplating, and manual polishing. The MMP-treated Cu–15Sn showed the best performance but the cost of the MMP treatment is more (Zhang et al., 2016). Especially for high-frequency antennas, conductor loss hampers the antenna's performance; hence the low-cost solution to tackle surface roughness needs to be investigated. Poor resolution and porosity in the structure are other basic limitations of current 3DP technologies.

To solve the issue of the lack of commercially available 3D printable materials with required electromagnetic properties (ex. loss tangent, dielectric constant), different methods to artificially control such parameters are already discussed in this chapter. High-quality conduction with 3D printed conductive polymer filament is also a challenge where ultrasonic wire embedding technique, an advanced nanocomposite polymer needs to be explored further for future antennas (Min Liang, 2014).

Nowadays, the supercycle of the commodity sector across the world leads to fluctuation in prices of various raw materials like copper, silver, etc. By inventing and adopting new technology like nanotechnology, 4D printing, etc., we have to get rid of price volatility in the world market.

In the case of nanocomposites, homogenous dispersion of nanoparticles is crucial for best performance. Uncured photopolymer in the Vat Polymerization process and clogging in nozzle occurs because of accumulation of filler particles. Addressing this issue is the need for time; the probable solution is already discussed. With the help of water transferring technology, the graphene-based antenna has recently been reported for IoT applications. Graphene ink is prepared by the liquid phase exfoliation method (Wang et al., 2019).

In the metallization or 3DP of metal, uniform coating is significant to achieve the required results and evade local stress concentrations, i.e., internal stresses. Uniform metallization and crystallinity are very important when designing an arbitrary shape antenna such as conformal antenna, generally placed on curvature surfaces like a missile, vehicles, airplanes, etc. Various metallization methods are explored in literature for 3D printed antennas but the various metallization process power handling capability of the metalized surface of the antenna needs to be evaluated (Hoel, Hellum, et al., 2016).

Recently aerosol jet process (AJP) has been successfully utilized to print nanomaterials like CNT, Graphene with high resolution on polycaprolactone (PLC) polymer to fabricate flexible, biodegradable, and conformal antennae. When AJP is utilized with AM process to converge with advanced printings will influence on-chip antenna manufacturing (Agarwala & Yeong, 2019; Wilkinson et al., 2019).

There is vast scope to explore advanced nanoparticle coating techniques, nanocomposite polymer printing technology, Multi-material i.e., metal and polymer simultaneous printing technology, and smart material like SMP for antenna fabrication.

3.9 SUMMARY

In this chapter, the overview of 3D AM techniques related to antenna design and implementation is described. The antenna categories i.e., dielectric, metal, and metal-coated dielectric printed from a 3DP perspective are discussed with example. 3DP antennas for various frequency bands with different 3DP techniques are explored here. The various material limitations and advantages concerning the conventionally manufactured antennas are discussed here. Current progress, opportunities, and challenges for 3DP of antennas are reviewed and discussed. The scope of SMP, which is less explored in antenna-related literature, is presented here. Nanomaterials and nanocomposite polymer for 3D printed antenna are discussed here. There are still substantial opportunities that need to be explored and specific challenges that need to be overcome to manufacture completely functional antennas via 3DP technology.

REFERENCES

Adeyeye, A. O., Bahr, R. A., & Tentzeris, M. M. (2019). 3D printed 2.45 GHz yagi-uda loop antenna utilizing microfluidic channels and liquid metal. *2019 IEEE International Symposium on Antennas and Propagation and USNC-URSI Radio Science Meeting, APSURSI 2019 - Proceedings*, 1983–1984. https://doi.org/10.1109/APUSNCURSINRSM.2019.8888346

Agarwala, S., & Yeong, W. Y. (2019). *Aerosol jet fabricated biodegradable antenna for bioelectronics application.* 1(1), 2–3. https://doi.org/10.18416/AMMM.2019.19.09S02T02

Ahmadloo, M., & Mousavi, P. (2013). A novel integrated dielectric-and-conductive ink 3D printing technique for fabrication of microwave devices. *IEEE MTT-S International Microwave Symposium Digest*, 29–31. https://doi.org/10.1109/MWSYM.2013.6697669

Akbari, M., Sydänheimo, L., Juuti, J., Vuorinen, J., & Ukkonen, L. (2014). Characterization of graphene-based inkjet printed samples on a flexible substrate for wireless sensing applications. *2014 IEEE RFID Technology and Applications Conference, RFID-TA 2014*, 135–139. https://doi.org/10.1109/RFID-TA.2014.6934215

Al Rashid, A., Khan, S. A., G. Al-Ghamdi, S., & Koç, M. (2021). Additive manufacturing of polymer nanocomposites: Needs and challenges in materials, processes, and applications. *Journal of Materials Research and Technology, 14*, 910–941. https://doi.org/10.1016/j.jmrt.2021.07.016

Andreas Gebhardt, U. A. M. (2012). *Understanding Additive Manufacturing Prototyping · Rapid Tooling. Rapid Manufacturing.* 4–7. https://doi.org/10.3139/9783446431621.FM

Bahr, R. A., Fang, Y., Su, W., Tehrani, B., Palazzi, V., & Tentzeris, M. M. (2017). Novel uniquely 3D printed intricate Voronoi and fractal 3D antennas. *IEEE MTT-S International Microwave Symposium Digest*, 1583–1586. https://doi.org/10.1109/MWSYM.2017.8058934

Balanis, C. A. (1992). Antenna Theory: A Review. *Proceedings of the IEEE*, 80(1), 7–23. https://doi.org/10.1109/5.119564

Bensoussan, H. (2016). The History of 3D Printing: From the 80s to Today. In *Sculpteo Blog.* https://www.sculpteo.com/en/3d-learning-hub/basics-of-3d-printing/the-history-of-3d-printing/

Bjorgaard, J., Hoyack, M., Huber, E., Mirzaee, M., Chang, Y. H., & Noghanian, S. (2018). Design and fabrication of antennas using 3D printing. *Progress In Electromagnetics Research C, 84* (February 2020), 119–134. https://doi.org/10.2528/pierc18011013

Busch, S. F., Weidenbach, M., Fey, M., Schäfer, F., Probst, T., & Koch, M. (2014). Optical Properties of 3D Printable Plastics in the THz Regime and their Application for 3D Printed THz Optics. *Journal of Infrared, Millimeter, and Terahertz Waves*, *35*(12), 993–997. https://doi.org/10.1007/s10762-014-0113-9

Cersoli, T., Barnawi, M., Johnson, K., Burden, E., Li, F., & Macdonald, E. (2022). *Recent Progress in Materials 4D Printed Shape Memory Polymers: Morphology and Fabrication of a Functional Antenna.* https://doi.org/10.21926/rpm.2202009

Chieh, J. C. S., Dick, B., Loui, S., & Rockway, J. D. (2014). Development of a ku-band corrugated conical horn using 3d print technology. *IEEE Antennas and Wireless Propagation Letters*, *13*, 201–204. https://doi.org/10.1109/LAWP.2014.2301169

Chietera, F. P., Colella, R., & Catarinucci, L. (2022). Dielectric resonators antennas potential unleashed by 3D printing technology: A practical application in the IoT framework. *Electronics (Switzerland)*, *11*(1). https://doi.org/10.3390/electronics11010064

Colella, R., Chietera, F. P., Catarinucci, L., Salmeron, J. F., Rivadeneyra, A., Carvajal, M. A., Palma, A. J., & Capitan-Vallvey, L. F. (2019). Fully 3D-Printed RFID Tags based on Printable Metallic Filament: Performance Comparison with other Fabrication Techniques. *Proceedings of the 2019 9th IEEE-APS Topical Conference on Antennas and Propagation in Wireless Communications, APWC 2019*, 253–257. https://doi.org/10.1109/APWC.2019.8870405

Dong, J., Huang, X., Muley, P., Wu, T., Barekati-Goudarzi, M., Tang, Z., Li, M., Lee, S., Boldor, D., & Wu, Q. (2020). Carbonized cellulose nanofibers as dielectric heat sources for microwave annealing 3D printed PLA composite. *Composites Part B: Engineering*, *184*, 107640. https://doi.org/10.1016/j.compositesb.2019.107640

Ehrmann, G., & Ehrmann, A. (2021). 3D printing of shape memory polymers. *Journal of Applied Polymer Science*, *138*(34), 1–11. https://doi.org/10.1002/app.50847

Fang, Y., Berrigan, J. D., Cai, Y., Marder, S. R., & Sandhage, K. H. (2012). Syntheses of nanostructured Cu- and Ni-based micro-assemblies with selectable 3D hierarchical biogenic morphologies. *Journal of Materials Chemistry*, *22*(4), 1305–1312. https://doi.org/10.1039/c1jm13884g

Farooqui, M. F., & Kishk, A. (2019). 3D -Printed Tunable Circularly Polarized Microstrip Patch Antenna. *IEEE Antennas and Wireless Propagation Letters*, *18*(7), 1429–1432. https://doi.org/10.1109/LAWP.2019.2919255

Farooqui, M. F., & Shamim, A. (2017). 3D Inkjet-Printed Helical Antenna with Integrated Lens. *IEEE Antennas and Wireless Propagation Letters*, *16*(c), 800–803. https://doi.org/10.1109/LAWP.2016.2604497

Fenn, A. J., Pippin, D. J., Lamb, C. M., Willwerth, F. G., Aumann, H. M., & Doane, J. P. (2016). 3D printed conformal array antenna: Simulations and measurements. *IEEE International Symposium on Phased Array Systems and Technology*, *0*, 5–8. https://doi.org/10.1109/ARRAY.2016.7832591

Gnanasekaran, K., Heijmans, T., Bennekom, S., Van, Woldhuis, H., & Wijnia, S. (2017). 3D printing of CNT- and graphene-based conductive polymer nanocomposites by fused deposition modeling. *Applied Materials Today*, *9*, 21–28. https://doi.org/10.1016/j.apmt.2017.04.003

Goesmann, H., & Feldmann, C. (2010). Nanoparticulate functional materials. *Angewandte Chemie - International Edition*, *49*(8), 1362–1395. https://doi.org/10.1002/anie.200903053

Gu, C., Gao, S., Fusco, V., Gibbons, G., Sanz-Izquierdo, B., Standaert, A., Reynaert, P., Bosch, W., Gadringer, M., Xu, R., & Yang, X. (2020). A D-Band 3D-Printed Antenna. *IEEE Transactions on Terahertz Science and Technology*, *10*(5), 433–442. https://doi.org/10.1109/TTHZ.2020.2986650

Hales, S. (2020). 3D printed nanomaterial-based electronic, biomedical, and bioelectronic devices. *Nanotechnology*, *3*.

Hasni, U., Green, R., Filippas, A. V., & Topsakal, E. (2019). One-step 3D-printing process for microwave patch antenna via conductive and dielectric filaments. *Microwave and Optical Technology Letters*, *61*(3), 734–740. https://doi.org/10.1002/mop.31607

Helena, D., Ramos, A., Varum, T., & Matos, J. N. (2020). Inexpensive 3D-Printed Radiating Horns for Customary Things in IoT Scenarios. *14th European Conference on Antennas and Propagation, EuCAP 2020*, 5–8. https://doi.org/10.23919/EuCAP48036.2020.9135972

Hoel, K. V., Hellum, T., & Kristoffersen, S. (2016). High power properties of 3D-printed antennas. *2016 IEEE Antennas and Propagation Society International Symposium, APSURSI 2016 - Proceedings*, *2*, 823–824. https://doi.org/10.1109/APS.2016.7696120

Hoel, K. V., Kristoffersen, S., Moen, J., Kjelgård, K. G., & Lande, T. S. (2016). Broadband antenna design using different 3D printing technologies and metallization processes. *2016 10th European Conference on Antennas and Propagation, EuCAP 2016*, 2–6. https://doi.org/10.1109/EuCAP.2016.7481620

Huang, Y., Bu, N., Duan, Y., Pan, Y., Liu, H., Yin, Z., & Xiong, Y. (2013). Electrohydrodynamic direct-writing. *Nanoscale*, *5*(24), 12007–12017. https://doi.org/10.1039/c3nr04329k

Huang, G. L., Zhou, S. G., & Chio, T. H. (2016). Lightweight perforated horn antenna enabled by 3D metal-direct-printing. *2016 IEEE Antennas and Propagation Society International Symposium, APSURSI 2016 - Proceedings*, 481–482. https://doi.org/10.1109/APS.2016.7695949

Khanna, P., Kaur, A., & Goyal, D. (2019). Algae-based metallic nanoparticles: Synthesis, characterization, and applications. *Journal of Microbiological Methods*, *163*(June), 105656. https://doi.org/10.1016/j.mimet.2019.105656

Kotze, K., & Gilmore, J. (2019). SLM 3D-Printed Horn Antenna for Satellite Communications at X-band. *Proceedings of the 2019 9th IEEE-APS Topical Conference on Antennas and Propagation in Wireless Communications, APWC 2019*, 148–153. https://doi.org/10.1109/APWC.2019.8870367

Kumar, P., Vaid, S., Singh, S., Kumar, J., Kumar, A., & Dwari, S. (2021). Design of 3D Printed Multi-Wavelength DRA. *IETE Technical Review (Institution of Electronics and Telecommunication Engineers, India)*, *38*(6), 662–671. https://doi.org/10.1080/02564602.2020.1819890

Lee, S. Y., Choo, M., Jung, S., & Hong, W. (2018). Optically transparent nano-patterned antennas: A review and future directions. *Applied Sciences (Switzerland)*, *8*(6), 1–13. https://doi.org/10.3390/app8060901

Lens, S. M. D., Applications, X., Belen, A., Mahouti, P., Filiz, G., & Tari, Ö. (2021). *Gain Enhancement of a Traditional Horn Antenna using 3D Printed*. *36*(2), 132–138.

Liang, M., Yu, X., Shemelya, C., Roberson, D., Macdonald, E., Wicker, R., & Xin, H. (2014). Electromagnetic materials of artificially controlled properties for 3D printing applications. *IEEE Antennas and Propagation Society, AP-S International Symposium (Digest)*, *1*, 227–228. https://doi.org/10.1109/APS.2014.6904445

Lomakin, K., Pavlenko, T., Ankenbrand, M., Sippel, M., Ringel, J., Scheetz, M., Klemm, T., Graf, D., Helmreich, K., Franke, J., & Gold, G. (2018). Evaluation and characterization of 3D printed pyramid horn antennas utilizing different deposition techniques for conductive material. *IEEE Transactions on Components, Packaging and Manufacturing Technology*, *8*(11), 1998–2006. https://doi.org/10.1109/TCPMT.2018.2871931

Lugo, D. C., Ramirez, R. A., Castro, J., Wang, J., & Weller, T. M. (2017). 3D printed multi-layer mm-wave dielectric rod antenna with enhanced gain. *2017 IEEE Antennas and Propagation Society International Symposium, Proceedings*, *2017-Janua*, 1247–1248. https://doi.org/10.1109/APUSNCURSINRSM.2017.8072666

Mazingue, G., Byrne, B., Romier, M., & Capet, N. (2020). 3D Printed Ceramic Antennas for Space Applications. *14th European Conference on Antennas and Propagation, EuCAP 2020*, 3–7. https://doi.org/10.23919/EuCAP48036.2020.9135312

Min Liang, C. S. (2014). Fabrication of microwave patch antenna using additive manufacturing technique. *The Acupuncture, 33*(3), 181–187. https://doi.org/10.13045/acupunct.2016045

Mirmozafari, M., Saeedi, S., Saeidi-Manesh, H., Zhang, G., & Sigmarsson, H. H. (2018). Direct 3D printing of nonplanar linear-dipole-phased array antennas. *IEEE Antennas and Wireless Propagation Letters, 17*(11), 2137–2140. https://doi.org/10.1109/LAWP.2018.2860463

Mirzaee, M., & Noghanian, S. (2017). 3D printed antenna using biocompatible dielectric material and graphene. *2017 IEEE Antennas and Propagation Society International Symposium, Proceedings, 2017-Janua*, 2543–2544. https://doi.org/10.1109/APUSNCURSINRSM.2017.8073314

Mohamadzade, B., Simorangkir, R. B. V. B., Hashmi, R. M., Gharaei, R., Lalbakhsh, A., Shrestha, S., Zhadobov, M., & Sauleau, R. (2021). A conformal, dynamic pattern-reconfigurable antenna using conductive textile-polymer composite. *IEEE Transactions on Antennas and Propagation, 69*(10), 6175–6184. https://doi.org/10.1109/TAP.2021.3069422

Poyanco, J. M., Pizarro, F., & Rajo-Iglesias, E. (2022). Cost-effective wideband dielectric planar lens antenna for millimeter-wave applications. *Scientific Reports, 12*(1), 1–10. https://doi.org/10.1038/s41598-022-07911-z

Ramade, C., Silvestre, S., Pascal-Delannoy, F., & Sorli, B. (2012). Thin-film HF RFID tags are deposited on paper by thermal evaporation. *International Journal of Radio Frequency Identification Technology and Applications, 4*(1), 49–66. https://doi.org/10.1504/IJRFITA.2012.044648

Reinhardt, A., Mobius-Labinski, M., Asmus, C., Bauereiss, A., & Hoft, M. (2019). Additive Manufacturing of 300 GHz Corrugated Horn Antennas. *IMWS-AMP 2019 - 2019 IEEE MTT-S International Microwave Workshop Series on Advanced Materials and Processes for RF and THz Applications, 1*, 40–42. https://doi.org/10.1109/IMWS-AMP.2019.8880123

Rodriguez, C., Avila, J., & Rumpf, R. C. (2016). *Ultra-Thin 3D Printed All-Dielectric Antenna. 64*(May), 117–123.

Rojas-Nastrucci, E. A., Nussbaum, J., Weller, T. M., & Crane, N. B. "Metallic 3D printed Ka-band pyramidal horn using binder jetting," *2016 IEEE MTT-S Latin America Microwave Conference (LAMC)*, 2016, pp. 1–3. https://doi.org/10.1109/LAMC.2016.7851297.

Rojas-Nastrucci, E. A., Nussbaum, J. T., Crane, N. B., & Weller, T. M. (2017). Ka-band characterization of binder jetting for 3-D printing of metallic rectangular waveguide circuits and antennas. *IEEE Transactions on Microwave Theory and Techniques, 65*(9), 3099–3108.

Sairam, S. K. K. V. S. S. S. S., & Singh, A. (2022). Graphene plasmonic nano-antenna for terahertz communication. *SN Applied Sciences, October 2021*. https://doi.org/10.1007/s42452-022-04986-1

Shamvedi, D., McCarthy, O. J., O'Donoghue, E., O'Leary, P., & Raghavendra, R. (2017). 3D metal printed sierpinski gasket antenna. *Proceedings of the 2017 19th International Conference on Electromagnetics in Advanced Applications, ICEAA 2017*, 633–636. https://doi.org/10.1109/ICEAA.2017.8065326

Sravani, P., & Rao, M. (2015). Design of 3D Antennas for 24 GHz ISM Band Applications. *Proceedings of the IEEE International Conference on VLSI Design, 2015-Febru*(February), 470–474. https://doi.org/10.1109/VLSID.2015.85

Toy, Y. C., Mahouti, P., Güneş, F., & Belen, M. A. (2017). Design and manufacturing of an X-band horn antenna using 3D printing technology. *Proceedings of 8th International Conference on Recent Advances in Space Technologies, RAST 2017*, 195–198. https://doi.org/10.1109/RAST.2017.8002988

Wang, S., Zhu, L., Li, Y., Zhang, G., Yang, J., Wang, J., & Wu, W. (2020). Radar Cross-Section Reduction of Helical Antenna by Replacing Metal with 3D Printed Zirconia Ceramic. *IEEE Antennas and Wireless Propagation Letters, 19*(2), 350–354. https://doi.org/10.1109/LAWP.2019.2962524

Wang, W., Ma, C., Zhang, X., Shen, J., & Hanagata, N. (2019). High-performance printable 2.4 GHz graphene-based antenna using water-transferring technology. *Science and Technology of Advanced Materials, 20*(1), 870–875. https://doi.org/10.1080/14686996.2019.1653741

Wang, X., Cheng, Y., & Dong, Y. (2022). Millimeter-Wave Dual-Polarized Metal Transmitarray Antenna with Wide Gain Bandwidth. *IEEE Antennas and Wireless Propagation Letters, 21*(2), 381–385. https://doi.org/10.1109/LAWP.2021.3132172

Wilkinson, N. J., Smith, M. A. A., Kay, R. W., & Harris, R. A. (2019). *A review of aerosol jet printing — a non-traditional hybrid process for micro-manufacturing.* 4599–4619.

Wu, J., Yuan, C., Ding, Z., Isakov, M., Mao, Y., Wang, T., Dunn, M. L., & Qi, H. J. (2016). Multi-shape active composites by 3D printing of digital shape memory polymers. *Scientific Reports, 6*(April), 1–11. https://doi.org/10.1038/srep24224

Xue, J., Wu, T., Dai, Y., & Xia, Y. (2019). Electrospinning and electrospun nanofibers: Methods, materials, and applications [Review-article]. *Chemical Reviews, 119*(8), 5298–5415. https://doi.org/10.1021/acs.chemrev.8b00593

Zhang, B., Linnér, P., Karnfelt, C., Tarn, P. L., Södervall, U. S. and Zirath, H. (2015). Attempt of the metallic 3D printing technology for millimeter-wave antenna implementations. Asia-Pacific Microwave Conference (APMC), 6, 1–3. https://doi.org/10.1109/APMC.2015.7413011.

Zhang, B., Zhan, Z., Cao, Y., Gulan, H., Linnér, P., Sun, J., Zwick, T., & Zirath, H. (2016). Metallic 3D Printed Antennas for Millimeter- and Submillimeter Wave Applications. *IEEE Transactions on Terahertz Science and Technology, 6*(4), 592–600. https://doi.org/10.1109/TTHZ.2016.2562508

Zhu, J., Yang, Y., McGloin, D., Liao, S., & Xue, Q. (2021). 3D Printed All-Dielectric Dual-Band Broadband Reflectarray With a Large Frequency Ratio. *IEEE Transactions on Antennas and Propagation, 69*(10), 7035–7040. https://doi.org/10.1109/TAP.2021.3076528

4 Polymer-Based 3D Printed Sensors, Actuators, and Antennas for Low-Cost Product Manufacturing

Ekta Thakur[1] and Isha Malhotra[2]
[1]Chandigarh University, Mohali, Punjab, India
[2]Global Institute of Technology and Management, Gurgaon, India

CONTENTS

DOI: 10.1201/9781003194224-4

4.1 INTRODUCTION

Miniaturized electronic gadgets have been in high demand for monitoring human body functions throughout the previous decade. As a result, wearable health management systems are becoming a popular study topic. Wearable technologies are becoming more popular since they can replace numerous medical tools when implanted in smart clothing (Vijayan et al., 2020). Researchers have created several tiny antennas with suitable performance for use with this technique (Hou, J. et al., 2019). However, to use wearable devices in biomedical applications, certain realistic antenna design requirements must be met, including small size, low weight, low power consumption, and a flexible structure. Microstrip antennas are becoming increasingly popular for wearable electronics that are positioned near the body. As a result, there has been an increased need for wearable antennas that are small and do not require any further setup for self-adapting wireless systems (Newman, P. 2020). It's worth noting that wireless body area networks are a prominent focus for wearable antennas (Wang, D. et al., 2018). As a result, printed circuit antennas have gained in popularity due to their unique qualities, such as their ability to meet the needs of wearable systems (Zhan, Y. et al., 2014). The cost-effectiveness, design simplicity, and biocompatibility of printed antennas make them ideal for wearable devices (Thakur E. et al., 2019). The ability to install printed circuit antennas on flexible and/or semiflexible substrates ensures their effectiveness against physical bending and twisting effects, which is one advantage of employing them with wearable devices (Liu, Y. et al., 2018). Apart from biomedical applications, a flexible antenna for harsh environments is one of the subjects of interest for Federal agencies, industry, and academia. Recent research (Li, R. et al., 2018, Gao, W. et al., 2020) has looked at the materials, construction, and applications of flexible and wearable antennas. Huang, S. et al., 2019 describe a practical strategy for developing and manufacturing wearable antennas that include nontextile and completely textile antennas. A printed monopole antenna and other flexible antenna manufacturing techniques were investigated in another study (Ghasemi, A. et al., 2008). A previous article (Singh, R. et al. 2017) provided an overview of materials and fabrication processes for wearable antennas in the very high-frequency (VHF) to the millimeter-wave band. Another study (Abhinav, K. et al., 2015) focused on wearable antenna materials and fabrication methods, as well as their applications, restrictions, and solutions. Current advancements in the arena of wearable ultra-wideband (UWB) antennae and their tenderness in wireless body-area network (WBAN) systems were emphasized in a previous study (Masihi, S. et al., 2020). Recent research (Corchia, L. et al., 2019) looked at dissimilar implanted antennas, their design necessities, and characteristics comparisons. It discusses a variety of flexible antennas rather than focusing on a single type, for example, wearables. Second, rather than focusing on the operating frequency, the authors consider antenna applications throughout a broad frequency range. Furthermore, the unique version of this editorial focuses on investigations conducted in the previous five years. The choice or design of antennas for wireless applications varies substantially depending on climatic parameters, transmission power, and frequency range (Wojkiewicz, et al., 2016). Furthermore, the antenna's characteristic is inclined by the material utilized, the construction method used, and the substrate parameters.

The paper looks at the latest research in conductive materials, substrates, fabrication techniques, and their various applications in flexible antennas. In addition, the obstacles and future directions in flexible antenna research are discussed.

4.2 FLEXIBLE ANTENNA MATERIAL SELECTION

To create flexible antennas several conducting materials and substrates are used. When choosing a substrate, take into account the dielectric characteristics, mechanical distortion acceptance miniaturization defenselessness, and environmental endurance. The antenna's characteristics are determined by the conductive material used, such as radiation efficiency (based on electrical conductivity) radiation pattern and gain, etc.

4.2.1 CONDUCTIVE MATERIALS

To attain high gain, efficiency, and bandwidth, high conductivity conductive patterns are important for wireless applications. Resistance to mechanical deformation is another desirable attribute of conductive materials. Nanoparticle (NP) inks are frequently utilized to build flexible antennas because of their high electrical conductivity. Copper NPs do not form oxide as easily as silver NP ink (Arif, A. et al., 2019). The use of copper-based NPs for flexible antennas has only been observed in very few investigations. NPs are not the only electro-textile materials used in flexible antennas; nickel- and silver-plated Ni/Ag fabrics, copper-coated nylon fabrics, and nonwoven conductive fabrics are also common. A previous article (Rmili, H. et al., 2006) reviewed different textile and non-textile conductive materials for evolving flexible antennas. Flexible antennas have been developed using adhesive Cu (Kaufmann, T. et al., 2012), Cu tapes (Wojkiewicz, et al., 2016), and Cu covering (Shin, K. et al., 2013) A design of antennas fabricated with dissimilar conducting materials can be found in Figure 4.1. Conductive polymers like polyaniline (PANI) (Lee, J. S. et al., 2015), polypyrrole (PPy) (Mo, L. et al.,2019), and poly (3,4-ethylene dioxythiophene) polystyrene sulfonate (PEDOT: PSS) (Ankireddy, K. et al., 2017), can be used to make wearable antennas. To increase the conductivity of polymers, carbon nanotubes (CNT)(Guerchouche, K. et al., 2017), grapheme (Zare, Y. et al., 2020), and carbon NPs (Hamouda, Z. et al., 2018), have been added (Figure 4.1). Graphene-based flexible antennas are attractive due to their great mechanical qualities and good electrical conductivity. Previous studies have used graphene materials such as paper (Ravindran, A. et al., 2018), graphene nanoflakes ink (Elmobarak et al., 2017), (Scidà, A. et al., 2018), graphene oxide ink (Leng, T. et al., 2016) and graphene NP ink (Zhou, X. et al., 2020)to create flexible antennas. Flexible antennas with good performance require conducting traces with excellent deformation stability and electrical conductivity (Zhou, et al., 2020). Doping is used to improve the conductivity of various bending conductive materials so that they can withstand the mechanical strain and deformation without losing performance (Shin, K. et al., 2011), Conductive silicon embedded with silver nanowires (Park, J. et al., 2019), fluorine rubber stained with silver (Thielens, A. et al., 2018), conductive polymers with CNTs [Li, W. et al., 2018], liquid metals in stretchy substrates (Salonen, P. et al., 2004), and even stretchable fabric itself (Locher, et al., 2006) are just a few examples.

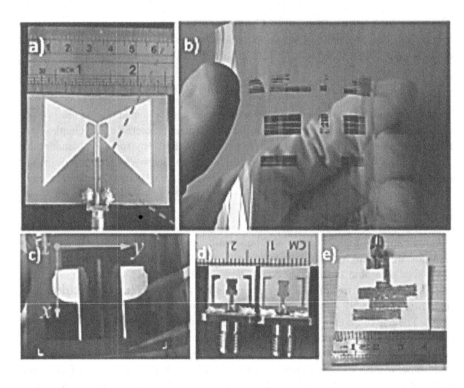

FIGURE 4.1 Layout of antenna: (a (PET) [67], (b) (PEN) [70], (c) Polyimide [59], (d) (LCP) [82], and (e) paper [80] substrates.

4.2.2 Substrate

Low insulator loss, less permittivity, less thermal expansion, and high thermal conductivity are all requirements for flexible antenna dielectric materials (Zhu, S. et al., 2009). The requirement for a high dielectric constant in tiny antennas is an exception to the rule. Three different kinds of substrates have been employed commonly in the production of wearable antennas: thin glass, metal halts, and polymer substrate (Scarpello, M.L. et al., 2012). Although thin glass can be bent, its inherent brittleness restricts its use. Metal foils can survive high temperatures and allow inorganic compounds to be coated on them, however, due to surface roughness and high material costs, their uses are limited (Ullah, M. et al., 2018). Plastic or polymer materials, such as the following, are the ideal prospects for flexible antenna applications: Tg materials: polyimide (PI) (Jilani, S. F. et al., 2019);. Polyethylene terephthalate (PET) and polyethylene naphthalate (PEN) are thermoplastics.

4.3 WEARABLE ANTENNA SENSOR FABRICATION METHODS

Fabrication procedures determine the precision and fabricating speed of cheap wearable antenna sensor devices. The traditional wearable production methods include wet-etching (Mohamadzade, B. et al., 2019)., screen printing (Mohamadzade, B. et al., 2019), inkjet printing (Liu, Q. et al., 2016), 3D printing, Chemical etching, and embroidery procedures (Rizwan, M. et al., 2019),.These techniques can be employed for antenna sensor manufacturing to assure durability, cheap cost, and excellent comfort for consumers in their everyday usage (Redwood, B. et al., 2017), (Cosker, et al., 2017, Mirzaee, M. et al., 2015). provides an intriguing overview of different manufacturing methods. The next sections go over a few of the manufacturing procedures.

4.3.1 SCREEN PRINTING

It is simple and lucrative technology; many electronics manufacturers use it to make a lightweight and wearable antenna sensor. It is also environmentally friendly because it is an additive process (Redwood, B. et al., 2017). Instead of using different thread densities and widths to hide the woven screen, the appropriate design is placed directly on the substrate and thermally annealed. In addition, screen-printing technology has a variety of disadvantages.

4.3.2 INKJET PRINTING

It is one of the more affordable printing methods (Mirzaee, M. et al., 2015) because it uses ink droplets as small as a few picoliters, this technology is capable of making an extremely precise design (Cosker. et al., 2017). Furthermore, this approach eliminates the need for masks by transmitting the design pattern directly to the substrate. Inkjet printing also ventures only ink droplets from the jet to the appropriate position, with no waste, making it one of the most cost-effective manufacturing methods.

In comparison to standard etching technology, which has been widely employed in the industry (Acti, T. et al., 2011); this is a distinct advantage. The mismatch of some types of conductive inks due to greater particle size and nozzle clogging are the main downsides of inkjet printing technology.

4.3.3 3D PRINTING

With a variety of commercially available printing materials and procedures, additive 3D printing techniques for wearable antennas have recently gained attention. It has several advantages, including the capacity to adjust the density of the printed object and fast fabrication of 3D structures using varied materials (Wang, Z. et al., 2014; Liu, Y. et al., 2010; Atzori, L. et al., 2010) (Yan, B. et al., 2014), (Kaim, V. et al., 2020). Fused deposition modeling (FDM) is the most widely used 3D printing method. The filament is connected to the printer's extrusion head, where it is melted by the heated nozzle. The melted material is then laid down by the printer in an accurate position, where it cools off and solidifies. The technique is repeated by layer-by-layer stacking of the portion (Kaim, V. et al., 2020).

4.3.4 CHEMICAL ETCHING

Chemical etching, which began in the 1960s as a part of the PCB industry, is the technique of manufacturing patterns by corrosively milling away a specified area

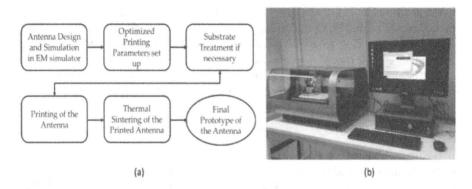

(a)   (b)

FIGURE 4.2   Fabrication process (a) Flowchart (b) Dimatix Materials Printer (Mirzaee, M. et al., 2017).

with photoresist and etchants. It is the finest choice among all other manufacturing techniques for creating complicated designs with high resolution precisely (Gao, Y. et al., 2011). Because their chemical properties alter when exposed to ultraviolet light, organic polymers are appropriate for photoresists. Because they offer a higher resolution than undesired resists, positive resists are more frequently utilized in photolithography-based antenna and RF circuits. This multilayer form of flexible antenna is made using the Physical Vapor Deposition (PVD) technology. Figure 4.2 depicts the fabrication steps. This approach allows for loops and grid arrangements. Four alternative fabrication processes, including photolithography (Singh, V.K. et al., 2017) were used to build a tiny epidermal Radio-frequency identification (RFID) antenna (2.5 cm × 5 cm). Figure 4.3 depicts the procedure. The antenna conductor is Au, and the PI substrate has a Ti/W adhesive layer. Although microfabrication has allowed for high-resolution throughput, the high expenses of clean rooms, photo masks, photolithography substances, and human properties will preclude the epidermal antenna from being cheap and throwaway.

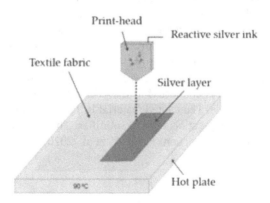

FIGURE 4.3   Inkjet printing (Cosker, et al., 2017).

4.3.5 EMBROIDERY

This method has developed to agreeing on a computer-assisted embroidery machine to stitch a digital picture or arrangement straight into a garment. Using specialized conductive threads used in embroidery manufacturing, the antenna can be blown up on the ignoble substrate textile fabric. Previously embroidering, it's vital to understand the belongings of the conductive threads that will be used, because once the conductive thread has been characterized, it's easier to develop ways to increase the antenna sensor's performance (Amit, S. et al., 2019). The conductive threads must have satisfactory confrontation and litheness to reduce unwanted breaks induced by strong stresses in the embroidery appliance. A computer-assisted embroidery machine can now sew a digital picture or arrangement right onto the fabric. The antenna sensor can be adorned on the base dielectric textile fabric using adequate conductive threads in the applique method of production. Understanding the qualities of the conductive threads that will be utilized is crucial because it will be easier to develop strategies to improve the antenna sensor's performance once the conductive thread has been defined (Soh, P.J. et al., 2011). The conductive threads must be resistant and flexible enough to prevent unintended breaks induced by high strains in the embroidery machine. Though embroidered antenna sensors are widely regarded as a superior alternative to traditional antennas in flexible electronics, they do have some disadvantages when compared to antenna sensors made of metallic materials. In the embroidery production method, the antenna sensor can be embroidered on the base substrate textile fabric using appropriate conductive threads. The resistance of the conductive yarn is significantly higher than that of copper materials. The cores are either carbon or nylon with silver plating. These yarn-based antenna sensors have a resistance that is orders of magnitude larger than metal or even printed antenna sensors.

4.3.6 COMPARISON

Figure 4.4 shows the embroidery procedure, which starts with a mock-up pattern that is typical of embroidering the antenna with the textile substrate. For example, embroidered antenna sensors on inlays are much more stretchable than metallic antenna sensors. Fine geometries are impossible to achieve due to the stretching impact and the limited high accuracy of the yarn stitches (Shaker, G. et al., 2011). The conductive yarn has a significantly higher resistance than metallic materials. Carbon or nylon cores with silver plating are used. The resistance of these yarn-based antenna sensors is orders of magnitude greater than that of metallic. Embroidery has a few advantages over other methods, due to embroidery machines becoming more popular in the industry. This technology lends itself better to mass-production clothing with embroidered antenna sensors. It is difficult to construct this form of building with Nora dell cloths or copper tape, currents in the fabric flow following the threads in needlework, attempting to make linear antenna sensors like dipoles ideal. It is extremely hard to construct this type of structure with Nora dell cloths or copper tape, currents in the fabric flow following the threads in needlework, attempting to make antenna sensors such

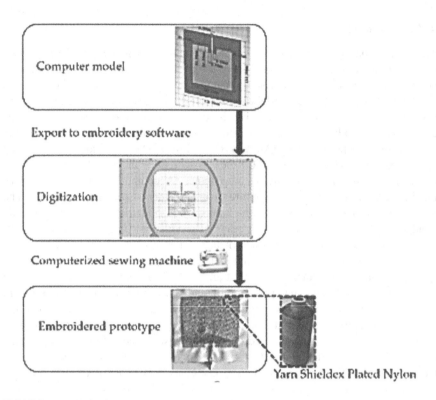

FIGURE 4.4   Embroidery process (Amit, S.. et al., 2019).

as dipoles ideal for this method. Glue isn't always mandatory to attend the cloth layers together and in embroidery, and it also allows computerized embroidery to create observable geometries (Cook, B.S. et al., 2012). The constructed antenna sensor improves the washability of the clothing.

4.4 APPLICATIONS OF FLEXIBLE ANTENNAS IN A VARIETY OF FREQUENCY BANDS

The most traditional forecasts predict that the global (IoT) sector would be worth USD 363.3 billion by 2025 (Hamza, S.M. et al., 2017). A significant portion of this market is made up of health monitoring and clinic treatment equipment, wearables, and car steering systems, among other things. To conform to curved surfaces and dynamic motions, the antenna used in these applications must be flexible, conformal, and stretchy. In addition to neutral applications, it plays an important role in the army and military. The majority of military devices are associated with a great ad-hoc network. Military personnel must transport a large amount of infrastructure, including sensors and wellness devices. As a result, it is lightweight and adaptable. As a result, flexible antennas are preferred in the military to lighten the load on soldiers. This article divides flexible antenna uses into two groups: below and above 12 GHz.

4.4.1 BELOW 12 GHz

A photo journal article bendable inkjet-printed RFID tag antenna for Ultra-high frequency (UHF) applications was described in a previously published study (Chen, S.J. et al., 2015). The antenna was developed for the global UHF band of 865–960 MHz and has an omnidirectional radiation pattern with a 4.57 m coverage. In a previous study (Chen, S.J. et al., 2015) a transitory tissue-type flexible RFID tag antenna with a range of up to 1.2 m was suggested. An additional study (Zhou, X. et al., 2020) described a flexible graphene nanoflakes antenna that was essentially a meandering line dipole. It had a radiation efficiency of 32 percent and a gain of 4 dBi and covered frequencies from 984 and 1052 MHz. A unique 3D printing skill to build an RFID tag antenna. A button-shaped RFID tag antenna was demonstrated by Yu, Y. et al., 2015 using 3D and inkjet printing methods. The maximum range of this antenna is 2.1 meters. A flexible 3D printed RFID tag antenna reached a maximum read range of 10.6 meters and a coverage of more than 7.4 meters under various stretching conditions. Consumer electronics and military applications all benefit from flexible antennas. In addition to UHF bands, the 2.45 GHz frequency is frequently used in wearables for industrial, scientific, and medical (ISM) applications. Antennas that are compact, lightweight, and long-lasting are ideal for this application. A flexible photo paper-based antenna operating at 2.33–2.53 GHz has been devised for intrabody telemedicine systems. A wearable textile logo antenna for military ISM band applications. Many additional flexible antennas for on-body usage have been documented in the ISM frequency band. The UWB antenna is another typical antenna. The UWB spectrum, which ranges from 3.1 to 10.6 GHz, was specified by the FCC in 2002 to meet the demand for faster communication speeds. Flexible wideband and UWB antennas are being studied for body-centric communications, a type of WBAN. UWB antennas have a compact electrical footprint, are inexpensive, have a low power spectrum density, and have a high data rate. The antenna is less likely to cause problems with other signals due to its lower spectral density (Chen, S.J. et al., 2015).

4.4.2 ABOVE 12 GHz

The textile-based UWB antenna has a negligible influence on the human body and can be employed for on-body applications (Yu, Y. et al., 2015 -Xiao, W. et al., 2017). This research (Simorangkir, R. B. et al., 2018), (Dwivedi, R.P. et al., 2020), used the first journal article on inkjet-printed UWB antenna. After that, a variety of conducting patch layouts was intended to recover the antenna's efficiency. In (Dwivedi, R.P. et al., 2020) paper, a polymer-based extensible antenna with high efficiency. The scientists employed PEDOT as the conducting material and sticky tape as the substrate in this experiment. In the literature, several polymer-based bendable UWB antennas have been described. Liquid crystal polymers (Cook, B.S.R.P. et al., 2013), polydimethylsiloxane (PDMS), graphene-assembled sheet, artificial magnetic conductor (AMC), PET, and polyamide are some of these materials. For example, flexible graphene antennas on a PI substrate working at 15 GHz produced huge bandwidth to accommodate higher speed communications for 5G applications (Tiercelin, N. et al., 2006).

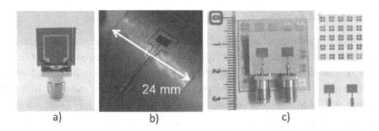

FIGURE 4.5 (a) Layout for 5G applications (Trajkovikj, J.. et al., 2012), (b) MMW antenna layout on PDMS substrate (El Atrash, et al., 2019), and (c) For wearable use EBG backed MMW MIMO antenna (Jilani, S.F.. et al., 2016).

An elastic, opalescent, and multiband mm-wave slotted antenna was developed, manufactured, and verified in a previous study (Iqbal, A. et al., 2019) using customized Ag NW ink inkjet printing elastic, opalescent, and the wideband mm-wave slotted antenna was developed, manufactured, and verified in a previous study (Iqbal, A. et al., 2019) using inkjet printing of customized Ag NW ink. The antenna had a very broad bandwidth, a radiation efficiency of 55%, and a maximum gain of 1.45 dB. In a previous study (Byungje, L. et al., 2002), 20 GHz coplanar waveguide (CPW) fed patch antennas on PET and Epson substrates were compared. To evaluate and compare this technology to the alternative, a micromachined patch antenna in the 60 GHz band was built on a PDMS dielectric in Figure 4.5. The researchers built an adaptable mm-wave antenna array with a bandwidth of 26–40 GHz, a peak gain of 11.35 dBi at 35 GHz, and strong gain characteristics of greater than 9 dBi across the entire Ka-band. In (Jilani, S.F. et al., 2016); proposed an EBG structured mm-wave MIMO antenna for wearable applications working at 24 GHz (ISM band) on a flexible Rogers substrate. Both free space and a bending human phantom were used to test the antenna configurations. Because of its simple shape and exceptional performance in having to bend and on-body worn conditions, the suggested antenna is well suitable for wearable electronics in the mm-wave range.

4.5 FLEXIBLE ANTENNA MINIATURIZATION

The ambition to join all electrical gadgets to the IoT has increased the demand for minor, more flexible antennas. Research into small antennas is becoming more prevalent. Researchers in this discipline face a tremendous problem in shrinking antennas to fit into small devices without sacrificing antenna performance factors such as -10 dB matching, gain, bandwidth, radiation pattern, and efficiency. Even though lowering antenna size is a tough and intimidating endeavor, scholars have devised a variety of inventive solutions. Various strategies for reducing antenna dimensions have been proposed in the literature. The strategies for reducing the form factor of flexible antennas are the topic of this paper.

The three primary categories are material-based, topology-based miniaturization, and the utilization of EBG structures. The use of high relative permittivity materials is the first method for dropping the flexible antenna dimension. The antenna's operational frequency is determined by its dielectric environment. Material-based, topology-based miniaturization, and the utilization of EBG structures are the three primary categories of approaches used. The use of high relative permittivity materials is the first method for dropping the flexible antenna size. The antenna's operational frequency is determined by its dielectric environment. The smaller the antenna becomes as the dielectric constant (k) increases. Many studies (Subramaniam, S. et al., 2014, Awan, W.A. et al., 2019; Mustafa, A. B. et al., 2019) use a similar methodology. A close-by filled high permittivity material was employed to minimize the antenna dimensions (Archevapanich, T. et al., 2014). The antenna properties alter as the shape, current density circulation, and electrical dimensions are changed. To ensure that an antenna has a specific property, optimization is required. Several twisting lines are used to expand the electrical length and miniaturize the flexible antenna. An asymmetric meander line was employed in previous work (Abdu, A. et al., 2018) to decrease the dimension of the wearable antenna while increasing the gain. A tiny dual-band flexible antenna was given in an earlier paper (Radoni´C. et al., 2012), which used a meander line to shrink the antenna and two-band characteristics. Because of the efficient use of space, the fractal antenna can deliver properties similar to a bigger antenna with a smaller size. Miniaturization was attained in habiliment electro textile antenna using Minkowski fractal geometry (Radoni´, C. et al., 2012). In (Waterhouse, R.B. et al., 1998); presented an ultra-thin bending antenna made up of rectangular fractal patches with a stub. When compared to the typical quadrilateral fractal patch, this fractal patch achieved 30% miniaturization.

An antenna's gain, radiation, and size were all controlled using a defective ground plane. This technique was used to create a small wearable antenna with a double flexible substrate (Iwasaki, H. 1996). Another technique to control the properties of the flexible antenna is to etch or print slots on the flexible substrate. The small, bending antenna with slots in the radiating patch was shown (Agneessens, S. et al., 2014). A Sierpinski carpet antenna on Hilbert slot design was presented in another study (Sievenpiper, D. et al., 1999) Cutting slots and shortening posts are two more frequent procedures (Thakur E., 2020-, Jaglan N. et al., 2018) (Raad H.R. et al., 2013) that have been employed in much research. A prominent method for reducing antenna size is the space-filling curve (SFC) (Raad H.R. et al., 2013). In a previous study (Abbasi, M.A.B. et al., 2017), the scholars included these downsizing methods with a 1/2 mode substrate integrity waveguide cavity to shrink the antenna even further. The potential for newly constructed EBG structures to decrease antenna physical structure whilst also maintaining radiation efficiency has captivated researchers' interest. Low-profile antennas were created using AMC and high impedance surfaces (HIS) (Yan, S. et al., 2014). This year saw the release of the first EBG-based wearable antenna (Velan, S. et al., 2015). Figure 4.6 shows the miniaturized antennas with EBG Structure. The antenna, despite its role as a wearable antenna, is not built on a flexible substrate. On the other hand, this is the first study to show how the EBG surface could be used to reduce antenna size. In a prior study (Abbasi, M.A.B. et al., 2017), the researchers combined these two downsizing techniques with a 1/2 mode substrate integrity waveguide cavity to further reduce the antenna's size. The possibility of newly built EBG structures to reduce antenna physical structure while maintaining radiation efficiency has piqued interest. AMCs and HIS were used to

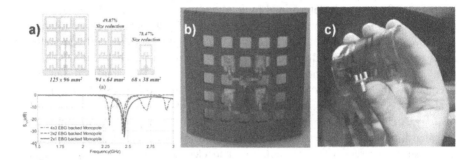

FIGURE 4.6 Miniaturized antennas. (a) EBG-Structure with S11 (Jin, Y. et al., 2016), Layout of PBG lattice (Suzan, M.M et al 2016), and (c) Jerusalem Cross (JC)-artificial magnetic conductor (AMC).

produce low-profile antennas (Yan, S. et al., 2014) The first EBG-based wearable antenna was other by this year (Velan, S. et al., 2015). Despite its role as a wearable antenna, the antenna is not constructed on a flexible substrate. On the other hand, this research is the first to demonstrate how the EBG surface can be used to decrease antenna size. The substrate for an AMC-based flexible M-shaped antenna for telemedicine applications is PI Kapton flexible material (Tsolis, et al., 2014). For telemedicine applications where the substrate was PI Kapton flexible material. The AMC structure aids in the isolation of antenna radiation from human flesh, as well as reducing the impedance mismatch produced by the user's body permittivity. Agneessens, S. et al., 2012 designed a textile wearable EBG antenna, with fleece fabric as the substrate. The EBG surface increased bandwidth by about 50% while reducing antenna size by around 30%. The scientists also looked at the antenna's stiffness under various bending situations and how it affected the impedance bandwidth. Following the success of this study, researchers expanded their research on EBG integration in antennas to produce tiny, high-performance antennas (Yan, S. et al., 2014, Velan. et al., 2015), Jiang, Z.H. et al., 2014, Ashyap, et al., 2018, Hong, J.H. et al., 2018, Hadarig. et al., 2013, R.C, Ashyap, et al., 2019, Salonen, P. et al., 2003.; Wang. et al., 2011, Scarpello. et al., 2011). Because of their periodic nature, photonic band gap (PBG) structures, another type of electromagnetic band-gap (EBG), can prohibit a wavelength from propagating. PBG is a three-dimensional structure made up of layered EBG layers. It's commonly made up of a tripod array and a multilayer metallic layer. Earlier work (Tsai, C.-L. et al., 2016) demonstrated the influence of PBG material on a traditional antenna system, as well as a method for reducing the antenna's size without sacrificing radiation efficiency, gain, or impedance bandwidth. The authors also used the proposed PBG material to show a new flexible antenna. Surface wave propagation was suppressed by using a PBG periodic structure in the conformal antenna and array (Alrawashdeh, R. et al., 2015)

4.6   DIFFERENT TYPES OF FLEXIBLE ANTENNA PERFORMANCE

The effectiveness of an antenna is determined by several factors, including the radiation element's conductivity, dielectric substrates, and various design concerns. The antenna's gain, efficiency, and bandwidth are all improved by using a highly conductive radiating element. For antenna performance, selecting the right dielectric material is crucial. With a larger loss tangent of the dielectric substrate, efficiency and gain are said to be diminished (Zhang, Y. et al., 2020), Furthermore, the antenna's bandwidth and resonant frequency are affected by dielectric permittivity (r). Antenna miniaturization with lower impedance bandwidth and minimal radiation losses is possible with a higher permittivity value (Zhang, Y. et al., 2020). Another element that can disturb efficiency, gain, bandwidth, and directivity is substrate thickness. When it comes to choosing the right substrate for a flexible antenna, thickness, performance, and flexibility are all factors to consider. In addition to the factors mentioned the previous section, antenna patch design, array arrangements, and power division transmission lines all have a significant impact on antenna characteristics. Patch elements come in a variety of shapes, including rectangular, square, and circular, with perturbed truncations (Ahmed, S. et al., 2015). These forms have an impact on polarization patterns, operating frequencies, $|S_{11}|$, and gain. Computer aided design (CAD) software in conjunction with an EM wave solver is required to iteratively analyze the designed antenna with an unconventional antenna. Table 4.1 compares the performance of various flexible antennas.

TABLE 4.1
Performance Comparison of Different Flexible Antenna

Reference	Dimension (mm³)	Substrate, Conductive material	Resonant Frequency	Gain Straight condition	Bending Condition	Application
(Lee, J.S,. et al., 2015)	39 × 39 × 0.508	Vinyl polymer-based flexible substrate, Cu	2.45 GHz	2.06 dBi.	−0.57 dBi	WBAN applications
(Zare, Y. et al., 2020)	60 × 60 × 0.110	Rogers RT/Duroid ® 5870, PANI/ MWCNTs	4.5 GHz.	5.18 dB	NA	C-band and future organic electronics applications
(Leng, T.. et al., 2016	46 × 45 mm²	Kapton PI, graphene flakes.	2 GHz.	2.3 dBi	NA	low-cost wireless communications applications.
(Trajkovikj, J.. et al., 2013)	40 × 38 × 0.135	PET, Ag NP	2.45 GHz	2.78 dBi	2.51 dBi	detection of brain stroke
(Khaleel, H.R. et al., 2012)	34 × 25 × 0.135	PET, Ag NP	NA	More than 3 dB	NA	ISM bands, UWB, WLAN band, WiMAX band, and 5G
Hasan, M.R. et al., 2020)	45 × 36 × 0.135	PET, Ag NP	900 MHz, 2.4 GHz.	6.74 and 16.24 dBi	NA	dual-band Wi-Fi and wearable devices

4.7 DEMANDS AND PROSPECTS FOR THE FUTURE OF FLEXIBLE ANTENNAS

Flexible wireless device research has recently gained a lot of focus because of its ability to meet the needs of biomedical use, systems for vehicle navigation, wearables, and other applications. The most important component in this system is an antenna, and it should be flexible and elastic to ensure device conformity. The first step toward achieving the goal is to use flexible materials to replace rigid substrates such as textiles, paper, or elastomeric polymers like PDMS (Amram B. et al., 2017), PEN, PET, and PI. As a result, selecting a suitable substrate is the very first obstacle in building a flexible antenna. Typical wearable substrates have low relative permittivity compared to standard substrates like FR4 or Rogers, which have a dielectric constant of 3–10 and a loss tangent of 0.001–0.02. Even while a low relative permittivity allows for greater bandwidth and radiation efficiency, it causes an issue when downsizing is required. The unequal thickness of flexible textile antennas is another issue to contend with. The electro textile substrate is vulnerable to crumbling and fluid absorption. An organic paper-based UWB antenna was previously described (Dey, et al., 2016). Although it is a low-profile antenna, the discontinuities and lack of resilience make it unsuitable for applications that demand a lot of bending and twisting. To tackle these issues, a polymer-based substrate is an ideal choice. An earlier study (Trajkovikj, J. et al., 2013).; looked at a small PI-based antenna that utilized Kapton PI material because of its low loss tangent (tan = 0.002) and physical and chemical flexibility for broadband frequency operation. This substrate can withstand temperatures up to 400°C and has a tensile strength of 165 MPa at 73°F, demonstrating the Kapton PI film's durability. PI and Kapton, which are translucent and bendable, are also inexpensive due to roll-to-roll mass manufacture. There are numerous different polymer-based designs available reported in the literature (Hasan, M.R. et al., 2020, Khaleel, H.R. et al., 2012). Excessive folding or twisting of the polymer-based antenna might cause micro-cracks in the substrate, which can be a concern. This will reduce the antenna's electrical conductivity and increase the chance of failure. Furthermore, polymers with low glass transition temperatures are unsuitable for high-temperature applications. Ceramic substrates can be a good alternative because they can endure high temperatures and are flexible (Govaert, F. et al., 2014). To address this constraint, very thin metallic nanowires can be embedded on the surface of elastomers like PDMS to make them highly conductive and flexible. It is not ideal for low-cost, flexible applications because of the manufacturing and design complexity. Liquid Metal (LM) is used in the microfluidic channel produced by elastomers instead of solid metal wire, the antenna will be reconfigurable, which is an exciting property of antenna that is desired in many applications. The microfluidic channel for the flexible antenna is made of PDMS, which is one of the most common commercial elastomers. Various liquid metals are poured into the channel to construct the antenna, including mercury, CNTs, Galinstan, gallium indium (GaIn), and eutectic gallium indium (EGaIn) (Sun, X. et al., 2013; Zhang, Y. et al., 2016, Babar, A.A. et al., 2012). In addition to PDMS, EcoFlex silicone rubber and TPU-based NinjaFlex are also used as elastomers for building microfluidic channels and are often 3D printed to achieve a specific pattern. Another issue in developing flexible antennas is finding appropriate conducting materials that can withstand various bending and twisting situations while maintaining a respectable resistance value without affecting the antenna's radiation efficiency. To obtain conductive substrates, many ways have been investigated, such as chemically altering fabric surfaces or physically mixing numerous conductive materials. Using the higher frequencies of the V (40–75 GHz) and W bands (75–110 GHz) is one way to minimize the antenna construction, regardless of the shape of the flexible antenna. A high-speed data connection is ensured by this frequent action. Materials with high dielectric constants are used to minimize the antenna's size. Most elastomeric materials have a low dielectric constant. Mixing the substrate with materials with high dielectric constants, such as BaxSr1xTiO3, BaTiO3, NdTiO3, MgCaTiO2, CNTs, and NPs, may help to boost this low value. The metamaterial-based bending antenna is a relatively great finding that has found its way into the commercial sector because of properties like lightweight, resilience, and reconfigurability.

4.8   FLEXIBLE ANTENNA FOR FUTURE WIRELESS SOLUTIONS

With the growing demand for wireless applications such as IoT, BAN, and bioelectronics, adjustable antennas are predictable to work across a wide range of frequencies. Antenna techniques include only one antenna, multiband antennas, and configurable antennas. Wireless LAN devices, for example, must operate in both the 2.4 and 5 GHz bands, which necessitates multiband architecture. According to the design, the antenna's properties should also correspond in bending scenarios. A low-priced inkjet-printed multiple-band antenna was presented in a previous study (Sun, X. et al., 2013). An original three-sided iterative aspect structure with CPW fed by printed on Kapton PI-based substrate material was used to realize multi-band operating with broad bandwidth. Bending antennas for emerging radio communication solutions are likely to perform in a wide range of frequencies due to the growing demand for wireless applications like the IoT, BAN, and medical and biological devices. A variety of antenna methods include single-band antennas, multiband antennas, and reconfigurable antennas. Multiple band design is frequently required; e.g., wireless LAN devices must operate in both the 2.4 and 5 GHz bands., The antenna's properties should remain consistent even when bent. Convex bending has no noteworthy resonance frequency change, but concave bending has a maximum 3 percent shift. Previous research established a flexible printed circuit (FPC)-based planar inverted-F antenna (PIFA) with the multi-band operation (Khan, S.H. et al., 2019). When the bending angle is less than 90 degrees, the antenna's characteristics remain constant. In (Abbasi, Q.H. et al., 2013) used wireless and satellite technology to create flexible and wearable antennas for IoT and WBAN applications. To prevent overcrowding in lower frequency bands, the antenna for satellite communication works in the C-band (4–8 GHz). The various categories of reconfigurable antennas are polarization, frequency, and pattern antennas. A major advantage of a reconfigurable antenna is the ability to swap bands built on the needs of the end-user. Prior work designed a versatile, spiral-shaped antenna that includes aeronautical radio navigation, WLAN, and WiMAX standards. The strip contains a lumped element for frequency reconfiguration, allowing the antenna to switch between multiple resonances.

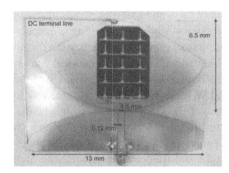

FIGURE 4.7 (a) Hybrid EM energy harvesting antenna (Collado, et al., 2013), (b) reconfigurable antenna with AMC surface (Yan, S. et al., 2018).

A paper (Collado, et al., 2013) and (Yan, S. et al., 2018) described polarization-based flexible, changeable antennas shown in Figure 4.7. The antenna is designed for use in biomedical applications that use WBAN and WiMAX standards, such as a remote monitoring system. Another article (Yan, S. et al., 2018) described a wearable pattern-recompose antenna. The inductor-loaded patch antenna's zero-order resonance and +1 resonance can be switched, resulting in two distinct patterns. The antenna was intended to function in the 2.4 GHz frequency range.

Using spectrum overlay with transmission power control, UWB technology provides for effective bandwidth usage. Using confining the program power, devices can work in 3.1–10.6 GHz with no instigating intrusion. Consequently, UWB technology is appalling for indoor and wearable applications. In a previous editorial (Patil, K.S, et al., 2019) a wearable band-rejected UWB antenna is suggested. To exclude interference from WLAN applications, a band notch feature is added for habiliment and interior UWB applications. No variation in antenna characteristics when twisted at various angles and can resist intense states. In (Wang, et al., 2019) a stretchy and see-through UWB antenna has been suggested. The antenna is comprised of a translucent copper material incorporated with PDMS. The antenna works from 2.2 GHz to 25 GHz. The negligible deviation is observed under folding. Aside from antennas' flexibility and ability to work with a wide range of wireless standards, some applications necessitate battery- and wire-free devices. Energy harvesting with a fixing antenna can be used to create self-contained devices. A rectenna collects radiofrequency energy from radio transmitters. Wireless power transfer was used in an RF-powered unleaded pacemaker in another investigation. A new wideband mathematical model was proposed by the authors (WBNM). The wideband numerical model was created using Tissue Simulating Liquid (TLS). A microstrip patch antenna was used to test the model both experimentally and analytically. In addition, a new metamaterial-based conformal implantable antenna working at 2.5 GHz has been designed. Wireless systems come with problems such as congestion, low bandwidth, and high latency. Millimeter-wave systems are expected to resolve these problems. With the upcoming 5G technology, these problems will be addressed and a greater data rate will be achieved thanks to a higher channel capacity and wider bandwidth. The design and shape of many smart devices are expected to be irregular and to be able to connect to the internet at high speeds. Antennas need to be able to stretch and conform when mounted on a conformal structure for this type of application. APET-based flexible T-shaped mm-wave antenna that covers a frequency range from 26 to 40 GHz. An antenna's bandwidth can be extended by combining multiple resonant points using a DGS. Due to heat sintering, the impedance matching at lower frequency ranges is not good, and the antenna performance under bending conditions has not been fully investigated experimentally. A PET-based Y-shaped antenna that is both flexible and transparent was created. In this study, the conductive material of the Ag HT type is used to ensure the conformity of the antenna. The resonant frequency and bandwidth of a Y-shaped antenna were also evaluated parametrically by changing the angle of the arms. From 27.71 to 24.60 GHz, the resonant frequency gradually decreases when the Y-shape structure is changed to a T-shape structure. The article claims that it can be dynamically tuned, which is not supported by the article. Another study (Kirtania, S.G. et al., 2020) proposes a wearable antenna that supports both 38 GHz and 60 GHz frequencies and is made of "ULTRALAM®3850HT," an ultra-flexible hybrid material. It is a four-sided patch antenna with six U-shaped slots. A commercial software package called HFSS and CST is used to verify the antenna response. This research has not conducted any experimental testing for the validity of the antenna response. To tackle the enormously growing and evolving IoT domain, 5G technology is on the horizon. The 5G network is expected to benefit a wide range of innovative solutions, including wearable devices robotics, self-driving cars, and so on. In addition to managing a large number of 'always connected' IoT., 5G will allow for the real-time management of massive amounts of data. Flexible antennas will be critical in the future implementation of a 5G wireless solution.

4.9 CONCLUSIONS

Flexible antennas are integrative, relating to mechanical engineering, materials science, and electrical engineering. The development of flexible electronic devices is only possible with flexible antennas. Flexible antennas are ideal for modern and future wireless communication and sensor applications due to the compact form factor, easy fabrication, and non-planar surfaces that can be placed. When fabricating antennas, materials are selected according to factors such as environmental factors, unified addition with inflexible and flexible devices, cost, and mass production characteristics of the manufacturing process. In the past, conductive patterns have typically been implemented using extremely conductive materials for example Ag NP inks, Cu tape or clad, conductive polymers, PDMS embedded conductive fiber, and graphene-based materials. The following substrates have been recommended as flexible dielectric: Kapton PI, PET, PEN, PANI, liquid crystal polymer, electrical, and journal. Flexible antennas are well suited to a range of applications above and below12 GHz. Various miniaturization techniques, as well as their challenges and limitations, have been discussed. For electromagnetic devices to be useful in medical applications, they need to be able to be surgically implanted and assimilable. The performance of flexible antennas has been investigated in terms of bends, strains, and proximity to the body. A symmetrically designed antenna or an increase in the bandwidth can assist account for distortion and other error-prone features.

REFERENCES

Abbasi, M. A. B.; Nikolaou, S. S.; Antoniades, M. A.; Nikolic Stevanovic, M.; Vryonides, P. Compact EBG-Backed Planar Monopole for BAN Wearable Applications. *IEEE Trans. Antennas Propag.* 2017, 65, 453–463.

Abbasi, Q. H.; Rehman, M. U.; Yang, X.; Alomainy, A.; Qaraqe, K.; Serpedin, E. (2013). Ultrawideband Band-Notched Flexible Antenna for Wearable Applications. *Antennas Wirel. Propag. Lett.*, 12, 1606–1609.

Abdu, A.; Zheng, H. -X.; Jabire, H. A.; Wang, M. (2018). CPW-Fed Flexible Monopole Antenna with H and Two Concentric C Slots on Textile Substrate, Backed by EBG for WBAN. *Int. J. RF Microw. Comput. Aided Eng.*, 28, e21505.

Abhinav K, V.; Rao R, V. K.; Karthik, P. S.; Singh, S. P. (2015) Copper conductive inks: Synthesis and utilization in flexible electronics. *RSC Adv.*, 5, 63985–64030.

Acti, T.; Zhang, S.; Chauraya, A.; Whittow, W.; Seager, R.; Dias, T. High-performance flexible fabric electronics for megahertz frequency communications. *In Proceedings of the 2011 Loughborough Antennas & Propagation Conference*, Loughborough, UK, 14–25 November; pp. 1–4.

Agneessens, S.; Bozzi, M.; Moro, R.; Rogier, H. (2012) Wearable textile antenna in substrate integrated waveguide technology. *Electron. Lett.*, 48, 985–987. 207.

Agneessens, S.; Rogier, H. Compact Half Diamond Dual-Band Textile HMSIW On-Body Antenna. (2014,) *IEEE Trans. Antennas Propag.* 62, 2374–2381.

Ahmed, M. I.; Ahmed, M. F.; Shaalan, A. -E. H. (2018). SAR Calculations of Novel Textile Dual-layer UWB Lotus Antenna for Astronauts Spacesuit. *Prog. Electromagn. Res. C*, 82, 135–144.

Ahmed, S.; Tahir, F. A.; Shamim, A.; Cheema, H. M. (2015). A Compact Kapton-Based Inkjet-Printed Multiband Antenna for Flexible Wireless Devices. *Antennas Wirel. Propag. Lett.*14, 1802–1805.

Alrawashdeh, R. S.; Yi, H.; Kod, M.; Abu Bakar Sajak, A. (2015). A Broadband Flexible Implantable Loop Antenna with Complementary Split Ring Resonators. *Antennas Wirel. Propag. Lett.*, 14, 1506–1509.

Amit, S.; Talasila, V.; Shastry, P. (2019) A Semi-Circular Slot Textile Antenna for UltraWideband Applications. *In Proceedings of the 2019 IEEE International Symposium on Antennas and Propagation and USNC-URSI Radio Science Meeting*, Atlanta, GA, USA, IEEE: Atlanta, GA, USA, 2019; pp. 249–250.

Amram Bengio, E.; Senic, D.; Taylor, L. W.; Tsentalovich, D. E.; Chen, P.; Holloway, C. L.; Babakhani, A.; Long, C. J.; Novotny, D. R.; Booth, J. C.; et al. (2017) High-efficiency carbon nanotube thread antennas. *Appl. Phys. Lett.*, 111, 163109.

Ankireddy, K.; Druffel, T.; Vunnam, S.; Filipi˘C, G.; Dharmadasa, R.; Amos, D.A. (2017). Seed Mediated Copper Nanoparticle Synthesis for Fabricating Oxidation-Free Interdigitated Electrodes Using Intense Pulse Light Sintering for Flexible Printed Chemical Sensors. *J. Mater. Chem. C*, 5, 11128–11137.

Archevapanich, T.; Lertwatechakul, M.; Rakluea, P.; Anantrasirichai, N.; Chutchavong, V. (2014) Ultra-Wideband Slot Antenna on Flexible Substrate for WLAN/WiMAX/UWB Applications. In *AsiaSim*.

Arif, A.; Zubair, M.; Ali, M.; Khan, M. U.; Mehmood, M. Q. (2019). A Compact, Low-Profile Fractal Antenna for Wearable On-Body WBAN Applications. *Antennas Wirel. Propag. Lett.*, 18, 981–985.

Ashyap, A. Y. I.; Zainal Abidin, Z.; Dahlan, S. H.; Majid, H. A.; Kamarudin, M. R.; Alomainy, A.; Abd-Alhameed, R. A.; Kosha, J. S.; Noras, J. M. (2018) Highly Efficient Wearable CPW Antenna Enabled by EBG-FSS Structure for Medical Body Area Network Applications. *IEEE Access*, 6, 77529–77541.

Ashyap, A. Y. I.; Zainal Abidin, Z.; Dahlan, S. H.; Majid, H. A.; Saleh, G. (2019) Metamaterial inspired fabric antenna for wearable applications. *Int. J. RF Microw. Comput. Aided Eng.*, 29, e21640.

Atrash, M. E.; Bassem, K.; Abdalla, M. A. (2017). A compact dual-band flexible CPW-fed antenna for wearable applications. *In Proceedings of the 2017 IEEE International Symposium on Antennas and Propagation & USNC/URSI National Radio Science Meeting*, San Diego, CA, USA, IEEE: San Diego, CA, USA, 2017; pp. 2463–2464.

Atzori, L.; Iera, A.; Morabito, G. (2010). The Internet of Things: A survey. *Comput. Netw.* 54, 2787–2805.

Awan, W. A.; Hussain, N.; Le, T. T. (2019). Ultra-thin flexible fractal antenna for 2.45 GHz application with wideband harmonic rejection. *AEU-Int. J. Electron. Commun.*, 110, 152851.

Babar, A. A.; Bjorninen, T.; Bhagavati, V. A.; Sydanheimo, L.; Kallio, P.; Ukkonen, L. (2012) Small and Flexible Metal Mountable Passive UHF RFID Tag on High-Dielectric Polymer-Ceramic Composite Substrate. *Antennas Wirel. Propag. Lett.*, 11, 1319–1322.

Byungje, L.; Harackiewicz, F. J. (2002) Miniature microstrip antenna with a partially filled high-permittivity substrate. *IEEE Trans. Antennas Propag.*, 50, 1160–1162.

Chen, S. J.; Kaufmann, T.; Shepherd, R.; Chivers, B.; Weng, B.; Vassallo, A.; Minett, A.; Fumeaux, C. (2015). A Compact, Highly Efficient, and Flexible Polymer Ultra-Wideband Antenna. *Antennas Wirel. Propag. Lett.*, 14, 1207–1210.

Collado, A.; Georgiadis, A. Conformal Hybrid Solar and Electromagnetic (EM) Energy Harvesting Rectenna. *IEEE Trans. Circuits Syst. I* 2013, 60, 2225–2234.

Cook, B. S.; Shamim, A. (2012,) Inkjet Printing of Novel Wideband and High Gain Antennas on Low-Cost Paper Substrate. *IEEE Trans. Antennas Propag.* 60, 4148–4156.

Cook, B.S.; Tehrani, B.; Cooper, J. R.; Tentzeris, M. M. (2013) Multilayer Inkjet Printing of Millimeter-Wave Proximity-Fed Patch Arrays on Flexible Substrates. *Antennas Wirel. Propag. Lett.*, 12, 1351–1354.

Corchia, L.; Monti, G.; Tarricone, L. (2019). Wearable Antennas: Nontextile versus Fully Textile Solutions. *IEEE Antennas Propag. Mag.*, 61, 71–83.

Cosker, M.; Lizzi, L.; Ferrero, F.; Staraj, R.; Ribero, J. -M. (2017) Realization of 3D Flexible Antennas Using Liquid Metal and Additive Printing Technologies. *Antennas Wirel. Propag. Lett.*, 16, 971–974.

Dey, A.; Guldiken, R.; Mumcu, G. (2016). Microfluidically Reconfigured Wideband Frequency-Tunable Liquid-Metal Monopole Antenna. *IEEE Trans. Antennas Propag.*, 64, 2572–2576.

Dwivedi, R. P.; Khan, M. Z.; Kommuri, U. K. (2020) UWB circular cross slot AMC design for radiation improvement of UWB antenna. *AEU-Int. J. Electron. Commun.*, 117, 153092.

El Atrash, M.; Abdalla, M. A.; Elhennawy, H. M. (2019) Gain enhancement of a compact thin flexible reflector-based asymmetric meander line antenna with low SAR. *IET Microw. Antennas Propag.* 13, 827–832.

Elmobarak Elobaid, H. A.; Abdul Rahim, S. K.; Himdi, M.; Castel, X.; Abedian Kasgari, M. A. (2017). Transparent and Flexible Polymer-Fabric Tissue UWB Antenna for Future Wireless Networks. *Antennas Wirel. Propag. Lett.*, 16, 1333–1336.

Gao, Y.; Zheng, Y.; Diao, S.; Toh, W. -D.; Ang, C. -W.; Je, M.; Heng, C.-H. (2011) Low-Power Ultrawideband Wireless Telemetry Transceiver for Medical Sensor Applications. *IEEE Trans. Biomed. Eng*, 58, 768–772.

Gao, W.; Zhu, Y.; Wang, Y.; Yuan, G.; Liu, J. -M. (2020) A review of flexible perovskite oxide ferroelectric films and their application. *J. Mater.*, 6, 1–16.

Ghasemi, A.; Sousa, E. S. (2008) Spectrum sensing in cognitive radio networks: Requirements, challenges, and design trade-offs. *IEEE Commun. Mag.*, 46, 32–39.

Govaert, F.; Vanneste, M. (2014) Preparation and Application of Conductive Textile Coatings Filled with Honeycomb Structured Carbon Nanotubes. *J. Nanomater.*, 2014, 651265.

Guerchouche, K.; Herth, E.; Calvet, L. E.; Roland, N.; Loyez, C. (2017) Conductive polymer-based antenna for wireless green sensors applications. *Microelectron. Eng.*, 182, 46–52.

Hadarig, R. C.; de Cos, M. E.; Las-Heras, F. (2013) UHF Dipole-AMC Combination for RFID Applications. *Antennas Wirel. Propag. Lett.* 2013, 12, 1041–1044.

Hamouda, Z.; Wojkiewicz, J. -L.; Pud, A. A.; Kone, L.; Bergheul, S.; Lasri, T. (2018).Magnetodielectric Nanocomposite Polymer-Based Dual-Band Flexible Antenna for Wearable Applications. *IEEE Trans. Antennas Propag.* 66, 3271–3277.

Hamza, S. M.; Tahir, F. A.; Cheema, H. M. (2017) A high-gain inkjet-printed UWB LPDA antenna on paper substrate. *Int. J. Microw. Wirel. Technol.*, 9, 931–937.

Hasan, M. R.; Riheen, M. A.; Sekhar, P.; Karacolak, T. (2020). Compact CPW-fed circular patch flexible antenna for super-wideband applications. *IET Microw. Antennas Propag.*, 14, 1069–1073.

Hasan, M. R.; Riheen, M. A.; Sekhar, P.; Karacolak, T. An Ink-jet Printed Flexible Monopole Antenna for Super Wideband Application. *Presented at the 2020 IEEE Texas Symposium Wireless & Microwave Circuits and Systems*, Waco, TX, USA, 26–28 May 2020.

Hertleer, C.; Rogier, H.; Vallozzi, L.; Van Langenhove, L. (2009) A Textile Antenna for Off-Body Communication Integrated Into Protective Clothing for Firefighters. *IEEE Trans. Antennas Propag.*, 57, 919–925.

Hong, J. H.; Chiu, C. -W.; Wang, H. -C. (2018) Design of Circularly Polarized Tag Antenna with Artificial Magnetic Conductor for on-body Applications. *Prog. Electromagn. Res. C*, 81, 89–99.

Hou, J.; Qu, L.; Shi, W. (2019) A survey on internet of things security from data perspectives. *Comput. Netw.*, 148, 295–306.

Huang, S.; Liu, Y.; Zhao, Y.; Ren, Z.; Guo, C. F. (2019) Flexible Electronics: Stretchable Electrodes and Their Future. *Adv. Funct. Mater.*, 29, 1805924.

Iqbal, A., *et al.*, (2019). Electromagnetic Bandgap Backed Millimeter-Wave MIMO Antenna for Wearable Applications, in *IEEE Access*, vol. 7, pp. 111135–111144.

Iqbal, A.; Basir, A.; Smida, A.; Mallat, N. K.; Elfergani, I.; Rodriguez, J.; Kim, S. (2019). Electromagnetic Bandgap Backed Millimeter-Wave MIMO Antenna for Wearable Applications. *IEEE Access*, 7, 111135–111144.

Iwasaki, H. (1996). A circularly polarized small-size microstrip antenna with a cross-slot. *IEEE Trans. Antennas Propag.*, 44, 1399–1401.

Jaglan N., Gupta, Samir Dev., Kanaujia, Binod., Srivastava, Shweta., and Thakur, Ekta. (2018). Triple Band Notched DG-CEBG Structure-Based UWB MIMO/Diversity Antenna, *Progress In Electromagnetics Research C*, 80, 21–37.

Jiang, Z. H.; Brocker, D. E.; Sieber, P. E.; Werner, D. H. (2014). A Compact, Low-Profile Metasurface-Enabled Antenna for Wearable Medical Body-Area Network Devices. *IEEE Trans. Antennas Propag.*, 62, 4021–4030.

Jilani, S.F.; Greinke, B.; Yang, H.; Alomainy, A. (2016) Flexible millimeter-wave frequency reconfigurable antenna for wearable applications in 5G networks. *In Proceedings of the 2016 URSI International Symposium on Electromagnetic Theory (EMTS)*, Espoo, Finland, 14–18 August 2016; IEEE: Espoo, Finland; pp. 846–848.

Jilani, S. F.; Abbasi, Q. H.; Alomainy, A. (2018) Inkjet-Printed Millimetre-Wave PET-Based Flexible Antenna for 5G Wireless Applications. *In Proceedings of the 2018 IEEE MTT-S International Microwave Workshop Series on 5G Hardware and System Technologies (IMWS-5G)*, Dublin, Ireland, 30–31; IEEE: Dublin, Ireland, 2018; pp. 1–3.

Jilani, S. F.; Munoz, M. O.; Abbasi, Q. H.; Alomainy, A. (2019) Millimeter-Wave Liquid Crystal Polymer Based Conformal Antenna Array for 5G Applications. *Antennas Wirel. Propag. Lett.*, 18, 84–88.

Jin, D.; Xiao, S.; Gao, S.; Tang, M.; Wang, B. (2010). Ultra wide-band slot antenna based on liquid crystal polymer material for millimeter-wave application. *In Proceedings of the 2010 IEEE International Conference on Ultra-Wideband*, Nanjing, China, 20–23 September 2010; IEEE: Nanjing, China, pp. 1–4.

Jin, Y.; Tak, J.; Choi, J. (2016) Broadband hybrid water antenna for ISM-band ingestible capsule endoscope systems. *In Proceedings of the 2016 International Workshop on Antenna Technology (iWAT)*, Cocoa Beach, FL, USA, 29 February–2 March 2016; IEEE: Cocoa Beach, FL, USA, pp. 82–85.

Kaim, V.; Kanaujia, B. K.; Kumar, S.; Choi, H. C.; Kim, K. W.; Rambabu, K. (2020) Ultra-Miniature Circularly Polarized CPW-Fed Implantable Antenna Design and its Validation for Biotelemetry Applications. *Sci. Rep.*, 10, 6795.

Kaufmann, T.; Verma, A.; Al-Sarawi, S. F.; Truong, V. -T.; Fumeaux, C. Comparison of two planar elliptical ultra-wideband PPy conductive polymer antennas. (2012), *In Proceedings of the 2012 IEEE International Symposium on Antennas and Propagation*, Chicago, IL, USA, 8–14; IEEE: Chicago, IL, USA, 2012; pp. 1–2.

Khaleel, H. R.; Al-Rizzo, H. M.; Rucker, D. G. (2012). Compact Polyimide-Based Antennas for Flexible Displays. *J. Display Technol.* 8, 91–97.

Khan, S. H.; Liu, T.; Zhang, L.; Razzaqi, A. A.; Khawaja, B. A. Design of flexible and wearable antenna for wireless and satellite-based IoT applications. *In Proceedings of the Tenth International Conference on Signal Processing Systems*, Singapore, 16–18 November 2018; Mao, K., Jiang, X., Eds.; SPIE: Singapore, 2019; Vol. 11071, pp. 220–227.

Kirtania, S. G.; Riheen, M. A.; Kim, S. U.; Sekhar, K.; Wisniewska, A.; Sekhar, P.K. (2020) Inkjet Printing on a New Flexible Ceramic Substrate for Internet of Things (IoT) Applications. *Micromachines*, 11, 841.

Lee, J. S.; Kim, M.; Oh, J.; Kim, J.; Cho, S.; Jun, J.; Jang, J. (2015). Platinum-Decorated Carbon Nanoparticle/Polyaniline Hybrid Paste for Flexible Wideband Dipole Tag-Antenna Application. *J. Mater. Chem. A*, 3, 7029–7035.

Leng, T.; Huang, X.; Chang, K.; Chen, J.; Abdalla, M. A.; Hu, Z. (2016,) Graphene Nanoflakes Printed Flexible Meandered-Line Dipole Antenna on Paper Substrate for Low-Cost RFID and Sensing Applications. *Antennas Wirel. Propag. Lett.* 15, 1565–1568.

Li, R.; Wang, L.; Yin, L. (2018) Materials and Devices for Biodegradable and Soft Biomedical Electronics. *Materials* 11, 2108.

Li, W.; Hei, Y.; Grubb, P. M.; Shi, X.; Chen, R.T. (2018) Compact Inkjet-Printed Flexible MIMO Antenna for UWB Applications. *IEEE Access*, 6, 50290–50298.

Liu, Q.; He, S.; Le, Tentzeris, M.M. (2016) Button-shaped radio-frequency identification tag combining three-dimensional and inkjet printing technologies. *IET Microw. Antennas Propag.*, 10, 737–741.

Liu, Y.; Wang, H.; Zhao, W.; Zhang, M.; Qin, H.; Xie, Y. Flexible (2018) Stretchable Sensors for Wearable Health Monitoring: Sensing Mechanisms, Materials, Fabrication Strategies and Features. *Sensors* 18, 645.

Liu, Y.; Xu, L.; Li, Y.; Ye, T. T. Textile-based embroidery-friendly RFID antenna design techniques. *In Proceedings of the 2019 IEEE International Conference on RFID (RFID)*, Pisa, Italy, 25–27 September 2019; pp. 1–6.

Locher, I.; Klemm, M.; Kirstein, T.; Troster, G. (2006). Design and Characterization of Purely Textile Patch Antennas. *IEEE Trans. Adv. Packag.*, 29, 777–788.

Masihi, S.; Panahi, M.; Maddipatla, D.; Bose, A. K.; Zhang, X.; Hanson, A. J.; Narakathu, B. B.; Bazuin, B. J.; Atashbar, M. Z. (2020). Development of a Flexible Tunable and Compact Microstrip Antenna via Laser-Assisted Patterning of Copper Film. *IEEE Sens. J.*, 20, 7579–7587.

Mirzaee, M.; Noghanian, S.; Wiest, L.; Chang, I. (2015) Developing flexible 3D-printed antenna using conductive ABS materials. *In Proceedings of the 2015 IEEE International Symposium on Antennas and Propagation & USNC/URSI National Radio Science Meeting*, Vancouver, BC, Canada, 19–24 July 2015; IEEE: Vancouver, BC, Canada, pp. 1308–1309.

Mo, L.; Guo, Z.; Wang, Z.; Yang, L.; Fang, Y.; Xin, Z.; Li, X.; Chen, Y.; Cao, M.; Zhang, Q.; et al. (2019). Nano-Silver Ink of High Conductivity and Low Sintering Temperature for Paper Electronics. *Nanoscale Res. Lett.*, 14, 197.

Mohamadzade, B.; Hashmi, R. M.; Simorangkir, R. B. V. B.; Gharaei, R.; Ur Rehman, S.; Abbasi, Q. H. (2019). Recent Advances in Fabrication Methods for Flexible Antennas in Wearable Devices: State of the Art. *Sensors*, 19, 2312.

Mustafa, A. B.; Rajendran, T. (2019) An Effective Design of Wearable Antenna with Double Flexible Substrates and Defected Ground Structure for Healthcare Monitoring System. *J. Med. Syst.*, 43, 186.

Newman, P. The Internet of Things 2020: Here's What over 400 IoT Decision-Makers Say about the Future of Enterprise Connectivity and How IoT Companies Can Use it to Grow Revenue.

Nikolaou, S.; Abbasi, M. A. B. (2016) Miniaturization of UWB Antennas on Organic Material. *Int. J. Antennas Propag.*, 5949254.

Park, J.; Park, S.; Yang, W.; Kam, D. G. (2019). Folded aperture coupled patch antenna fabricated on FPC with vertically polarized end-fire radiation for fifth-generation millimeter-wave massive MIMO systems. *IET Microw. Antennas Propag.* 13, 1660–1663.

Patil, K. S.; Rufus, E. (2019) A review on antennas for biomedical implants used for IoT-based health care. *Sens. Rev.*, 40, 273–280.

Raad, H. R.; Abbosh, A. I.; Al-Rizzo, H. M.; Rucker, D. G. (2013) Flexible and Compact AMC Based Antenna for Telemedicine Applications. *IEEE Trans. Antennas Propag.*, 61, 524–531

Radoni´C, V.; Palmer, K.; Stojanovi´C, G.; Crnojevi´c-Bengin, V. (2012) Flexible Sierpinski Carpet Fractal Antenna on a Hilbert Slot Patterned Ground. *Int. J. Antennas Propag*

Ravindran, A.; Feng, C.; Huang, S.; Wang, Y.; Zhao, Z.; Yang, J. (2018). Effects of Graphene Nanoplatelet Size and Surface Area on the AC Electrical Conductivity and Dielectric Constant of Epoxy Nanocomposites. *Polymers*, 10, 477.

Redwood, B.; Schöffer, F.; Garret, B. (2017) The 3D Printing Handbook: Technologies, Design, and Applications; 3D HUBS: Amsterdam, The Netherlands.

Rida, A.; Li, Y.; Vyas, R.; Tentzeris, M. M. (2009) Conductive Inkjet-Printed Antennas on Flexible Low-Cost Paper-Based Substrates for RFID and WSN applications. *IEEE Antennas Propag. Mag.* 2009, 51, 13–23.

Rizwan, M.; Khan, M. W. A.; Sydanheimo, L.; Virkki, J.; Ukkonen, L. (2017) Flexible and Stretchable Brush-Painted Wearable Antenna on a Three-Dimensional (3D) Printed Substrate. *Antennas Wirel. Propag. Lett.*, 16, 3108–3112.

Rmili, H.; Miane, J. -L.; Zangar, H.; Olinga, T. (2006), Design of microstrip-fed proximity-coupled conducting-polymer patch antenna. *Microw. Opt. Technol. Lett.* 48, 655–660.

Saeed, S. M.; Balanis, C. A.; Birtcher, C. R.; Durgun, A. C.; Shaman, H. N. (2017) Wearable Flexible Reconfigurable Antenna Integrated with Artificial Magnetic Conductor. *Antennas Wirel. Propag. Lett.*, 16, 2396–2399.

Salonen, P. (2001), A low-cost 2.45 GHz photonic band-gap patch antenna for wearable systems. *In Proceedings of the 11th International Conference on Antennas and Propagation (ICAP 2001)*, Manchester, UK, 17–20 April 2001; IEEE: Manchester, UK, 2001; Volume, pp. 719–723.

Salonen, P.; Fan, Y.; Rahmat-Samii, Y.; Kivikoski, M. W (2004). EBGA-wearable electromagnetic band-gap antenna. *In Proceedings of the IEEE Antennas and Propagation Society Symposium, Monterey*, CA, USA, 20–25 June 2004; IEEE: Monterey, CA, USA, 2004; 1, pp. 451–454.

Salonen, P.; Rahmat-Samii, Y.; Schaffrath, M.; Kivikoski, M. (2004) Effect of textile materials on wearable antenna performance: A case study of GPS antennas. *In Proceedings of the IEEE Antennas and Propagation Society Symposium, Monterey*, CA, USA, 20–25; *IEEE: Monterey*, CA, USA, Volume 1, pp. 459–462.

Salonen, P.; Sydänheimo, L.; Keskilammi, M. (2003) Antenna Miniaturization Using Flexible PBG Materials. *In Proceedings of the 2003 International Electronic Packaging Technical Conference and Exhibition*, Maui, HI, USA, 6–11 July 2003; ASMEDC: Maui, HI, USA, Volume 2, pp. 1–5.

Scarpello, M. L.; Kazani, I.; Hertleer, C.; Rogier, H.; Vande Ginste, D. (2012) Stability and Efficiency of Screen-Printed Wearable and Washable Antennas. *Antennas Wirel. Propag. Lett.*, 11, 838–841.

Scarpello, M. L.; Kurup, D.; Rogier, H.; Vande Ginste, D.; Axisa, F.; Vanfleteren, J.; Joseph, W.; Martens, L.; Vermeeren, G. (2011) Design of an Implantable Slot Dipole Conformal Flexible Antenna for Biomedical Applications. *IEEE Trans. Antennas Propag.* 59, 3556–3564.

Scidà, A.; Haque, S.; Treossi, E.; Robinson, A.; Smerzi, S.; Ravesi, S.; Borini, S.; Palermo, V. (2018). Application of graphene-based flexible antennas in consumer electronic devices. *Mater. Today*, 21, 223–230.

Sethi, P.; Sarangi, S. R. (2017). Internet of Things: Architectures, Protocols, and Applications. *J. Electr. Comput. Eng.,* 2017, 1–25.

Shaker, G.; Safavi-Naeini, S.; Sangary, N.; Tentzeris, M. M. (2011) Inkjet Printing of Ultrawideband (UWB) Antennas on Paper-Based Substrates. *Antennas Wirel. Propag. Lett.*, 10, 111–114.

Shin, K. -Y.; Cho, S.; Jang, J. (2013). Graphene/Polyaniline/Poly(4-styrenesulfonate) Hybrid Film with Uniform Surface Resistance and Its Flexible Dipole Tag Antenna Application. *Small*, 9, 3792–3798.

Shin, K. -Y.; Hong, J. -Y.; Jang, J. (2011) Micropatterning of Graphene Sheets by Inkjet Printing and Its Wideband Dipole-Antenna Application. *Adv. Mater.*, 23, 2113–2118.

Sievenpiper, D.; Lijun, Z.; Broas, R. F. J.; Alexopolous, N. G.; Yablonovitch, E. (1999). Highimpedance Electromagnetic Surfaces with a Forbidden Frequency Band. *IEEE Trans. Microw. Theory Tech89* 47, 2059–2074.

Simorangkir, R. B. V. B.; Kiourti, A.; Esselle, K. P. (2018) UWB Wearable Antenna with a Full Ground Plane Based on PDMS-Embedded Conductive Fabric. *Antennas Wirel. Propag. Lett.*, 17, 493–496.

Singh, R.; Singh, E.; Nalwa, H. S. (2017) Inkjet-printed nanomaterial-based flexible radio frequency identification (RFID) tag sensors for the internet of nano things. *RSC Adv.*, 7, 48597–48630.

Singh, V. K.; Dhupkariya, S.; Bangari, N. (2017) Wearable Ultra Wide Dual Band Flexible Textile Antenna for WiMax/WLAN Application. *Wirel. Pers. Commun.*, 95, 1075–1086.

Soh, P. J.; Vandenbosch, G. A. E.; Higuera-Oro, J. Design and evaluation of flexible CPW-fed Ultra-Wideband (UWB) textile antennas. *In Proceedings of the 2011 IEEE International RF & Microwave Conference*, Seremban, Malaysia, 12–14 December 2011; IEEE: Seremban, Malaysia, 2011; pp. 133–136.

Subramaniam, S.; Dhar, S.; Patra, K.; Gupta, B.; Osman, L.; Zeouga, K.; Gharsallah, A. (2014) Miniaturization of wearable electro-textile antennas using Minkowski fractal geometry. *In Proceedings of the 2014 IEEE Antennas and Propagation Society International Symposium* (APSURSI), Memphis, TN, USA, IEEE: Memphis, TN, USA, 2014; pp. 309–310.

Sun, X.; Sun, H.; Li, H.; Peng, H. (2013). Developing Polymer Composite Materials: Carbon Nanotubes or Graphene? *Adv. Mater.*, 25, 5153–5176.

Suzan, M. M.; Haneda, K.; Icheln, C.; Khatun, A.; Takizawa, K. (2016) An ultrawide band conformal loop antenna for ingestible capsule endoscope system. *In Proceedings of the 2016 10th European Conference on Antennas and Propagation (EuCAP)*, Davos, Switzerland, 10–15 April 2016; IEEE: Davos, Switzerland; pp. 1–5.

Thakur, E., Jaglan, Naveen., & Gupta, Samir Dev. (2019). Design of Compact UWB MIMO Antenna with Enhanced Bandwidth. *Progress In Electromagnetics Research C*, 97, 83–94,

Thakur E., Jaglan, Naveen., Gupta, Samir Dev., & Kanaujia, Binod (2019). A Compact Notched UWB MIMO Antenna with Enhanced Performance, *Progress In Electromagnetics Research C*, 91, 39–53.

Thakur, Ekta., Jaglan, Naveen., & Gupta, Samir Dev (2020) Design of compact triple band-notched UWB MIMO antenna with TVC-EBG structure, *Journal of Electromagnetic Waves and Applications*, 34:11, 1601–1615. doi:10.1080/092050 71.2020.1775136.

Thakur, E., Jaglan, N., & Gupta, S. Dev (2022). Miniaturized four-port UWB MIMO antennas with triple-band rejection using single EBG structures. *International Journal of Microwave and Wireless Technologies*, 14(2), 185–193. doi:10.1017/S1759078721000325

Thielens, A.; Deckman, I.; Aminzadeh, R.; Arias, A. C.; Rabaey, J. M. (2018). Fabrication and Characterization of Flexible Spray-Coated Antennas. *IEEE Access*, 6, 62050–62061.

Tiercelin, N.; Coquet, P.; Sauleau, R.; Senez, V.; Fujita, H. (2006) Polydimethylsiloxane membranes for millimeter-wave planar ultra-flexible antennas. *J. Micromech. Microeng.*, 16, 2389–2395.

Trajkovikj, J.; Zurcher, J. -F.; Skrivervik, A. K. (2012) Soft and flexible antennas on permittivity adjustable PDMS substrates. *In Proceedings of the 2012 Loughborough Antennas & Propagation Conference (LAPC)*, Loughborough, UK, 12–13 November 2012; IEEE: Loughborough, UK, 2012; pp. 1–4.

Trajkovikj, J.; Zurcher, J. -F.; Skrivervik, A. K., PDMS, (2013) A Robust Casing for Flexible W-BAN Antennas [EurAAP Corner]. IEEE Antennas Propag. Mag., 55, 287–297.

Tsai, C. -L.; Chen, K. -W.; Yang, C. -L. (2016). Implantable Wideband Low-Specific-Absorption-Rate Antenna on a Thin Flexible Substrate. *Antennas Wirel. Propag. Lett.*, 15, 1048–1052.

Tsolis, A.; Whittow, W.; Alexandridis, A.; Vardaxoglou, J. (2014). Embroidery and Related Manufacturing Techniques for Wearable Antennas: Challenges and Opportunities. *Electronics*, 3, 314–338.

Ullah, M.; Islam, M.; Alam, T.; Ashraf, F. (2018), Paper-Based Flexible Antenna for Wearable Telemedicine Applications at 2.4 GHz ISM Band. *Sensors*, 18, 4214.

Velan, S.; Sundarsingh, E. F.; Kanagasabai, M.; Sarma, A. K.; Raviteja, C.; Sivasamy, R.; Pakkathillam, J. K. (2015) Dual-Band EBG Integrated Monopole Antenna Deploying Fractal Geometry for Wearable Applications. *Antennas Wirel. Propag. Lett.*, 14, 249–252.

Velan, S.; Sundarsingh, E. F.; Kanagasabai, M.; Sarma, A. K.; Raviteja, C.; Sivasamy, R.; Pakkathillam, J. K. (2015). Dual-Band EBG Integrated Monopole Antenna Deploying Fractal Geometry for Wearable Applications. *Antennas Wirel. Propag. Lett.* 2015, 14, 249–252.

Venkateswara Rao, M.; Madhav, B. T. P.; Anilkumar, T.; Prudhvinadh, B. (2020) Circularly polarized flexible antenna on liquid crystal polymer substrate material with metamaterial loading. *Microw. Opt. Technol. Lett.*, 62, 866–874.

Vijayan, V.; Connolly, J. P., Condell, J., McKelvey, N., Gardiner, P. (2021) Review of wearable devices and data collection considerations for connected health. *Sensors (Basel)*, 2021 Aug 19, 21(16), 5589. doi:10.3390/s21165589. PMID: 34451032; PMCID: PMC8402237.

Wang, D.; Chen, D.; Song, B.; Guizani, N.; Yu, X.; Du, X. (2018) From IoT to 5G I-IoT: The Next Generation IoT-Based Intelligent Algorithms and 5G Technologies. *IEEE Commun. Mag.*, 56, 114–120.

Wang, W.; Ma, C.; Zhang, X.; Shen, J.; Hanagata, N.; Huangfu, J.; Xu, M.2019) High-performance printable 2.4 GHz graphene-based antenna using water-transferring technology. *Sci. Technol. Adv. Mater.*, 20, 870–875.

Wang, X.; Zhang, M.; Wang, S. -J.(2011). Practicability Analysis and Application of PBG Structures on Cylindrical Conformal Microstrip Antenna and Array. *Prog. Electromagn. Res.*, 115, 495–507.

Wang, Z.; Lee, L. Z.; Psychoudakis, D.; Volakis, J. L. (2014) Embroidered multiband body-worn antenna for GSM/PCS/WLAN communications. *IEEE Trans. Antennas Propag.*, 62, 3321–3329.

Waterhouse, R. B.; Targonski, S. D.; Kokotoff, D. M. (1998) Design, and performance of small printed antennas. *IEEE Trans. Antennas Propag.*, 46, 1629–1633.

Rao, S. K., & Prasad, R. (2018). Impact of 5G technologies on industry 4.0. *Wireless Personal Commun.*, 100(1), 145–159..

Wojkiewicz, J. -L.; Alexander, P.; Bergheul, S.; Belkacem, B.; Lasri, T.; Zahir, H.; Kone, L. (2016) Design fabrication and characterization of polyaniline and multiwall carbon nanotubes composites-based patch antenna. *IET Microw. Antennas Propag.*, 10, 88–93.

Xiao, W.; Mei, T.; Lan, Y.; Wu, Y.; Xu, R.; Xu, Y. (2017) Triple band-notched UWB monopole antenna on ultra-thin liquid crystal polymer-based on ESCSRR. *Electron. Lett.*, 53, 57–58.

Yadav, A.; Kumar Singh, V.; Kumar Bhoi, A.; Marques, G.; Garcia-Zapirain, B.; de la Torre Díez, I. (2020). Wireless Body Area Networks: UWB Wearable Textile Antenna for Telemedicine and Mobile Health Systems. *Micromachines*, 11, 558.

Yan, B.; Zhang, J.; Wang, H.; Shi, Y.; Zong, W. (2019). A Flexible Wearable Antenna Operating at ISM band. *In Proceedings of the 2019 International Workshop on Electromagnetics: Applications and Student Innovation Competition (iWEM)*, Qingdao, China, 18–20 September; IEEE: Qingdao, China, 2019; pp. 1–2.

Yan, S.; Soh, P. J.; Vandenbosch, G. A. E. (2014) Low-Profile Dual-Band Textile Antenna with Artificial Magnetic Conductor Plane. *IEEE Trans. Antennas Propag.*, 62, 6487–6490.

Yan, S.; Soh, P. J.; Vandenbosch, G. A. E. (2014). Low-Profile Dual-Band Textile Antenna with Artificial Magnetic Conductor Plane. *IEEE Trans. Antennas Propag.*, 62, 6487–6490.

Yan, S.; Soh, P. J.; Vandenbosch, G. A. E. (2018). Wearable Ultrawideband Technology—A Review of Ultrawideband Antennas, Propagation Channels, and Applications in Wireless Body Area Networks. *IEEE Access*, 6, 42177–42185.

Yu, Y.; Jin, P.; Ding, K.; Zhang, M. (2015) A flexible UWB CPW-fed antenna on liquid crystal polymer substrate. *In Proceedings of the 2015 Asia-Pacific Microwave Conference (APMC)*, Nanjing, China, 6–9 December 2015; IEEE: Nanjing, China, pp. 1–3.

Zare, Y.; Rhee, K. Y. (2020) Calculation of the Electrical Conductivity of Polymer Nanocomposites Assuming the Interphase Layer Surrounding Carbon Nanotubes. *Polymers*, 12, 404.

Zhan, Y.; Mei, Y.; Zheng, L. (2014) Materials capability and device performance in flexible electronics for the Internet of Things. *J. Mater. Chem. C*, 2, 1220–1232.

Zhang, Y.; Li, S.; Yang, Z.; Qu, X.; Zong, W. A coplanar waveguide-fed flexible antenna for ultra-wideband applications. *Int. J. RF Microw. Comput. Aided Eng.* 2020, 30, e22258.

Zhang, Y.; Shi, S.; Martin, R. D.; Prather, D. W. (2016) High-Gain Linearly Tapered Antipodal Slot Antenna on LCP Substrate at E- and W-Bands. *Antennas Wirel. Propag. Lett.*, 15, 1357–1360.

Zhou, X.; Leng, T.; Pan, K.; Abdalla, M. A.; Hu, Z. (2020). Graphene Printed Flexible and Conformal Array Antenna on Paper Substrate for 5.8 GHz Wireless Communications. In *Proceedings of the 2020 14th European Conference on Antennas and Propagation (EuCAP)*, Copenhagen, Denmark, 15–20 March 2020; IEEE: Copenhagen, Denmark, pp. 1–4.

Zhu, S.; Langley, R. (2009) Dual-Band Wearable Textile Antenna on an EBG Substrate. *IEEE Trans. Antennas Propag*, 57, 926–935.

5 Lab-Scale Solutions for Sensors, Actuators, and Antennas
Economic and Environmental Justification

Rupinder Singh, Abhishek Barwar, and Anish Das
National Institute of Technical Teachers Training
and Research, Chandigarh, India

CONTENTS

5.1 INTRODUCTION

With the rise in demand for the inclusion of intelligence in the field of electronic industries every year, conventional electronic device manufacturing techniques report a huge environmental impact (Wiklund et al., 2021). Additive manufacturing (AM) proves to be an effective method to manufacture such devices along with utilizing the electronic waste by consuming it in the form of feedstock filament or inks

DOI: 10.1201/9781003194224-5

(Maddipatla et al., 2020). AM with no need for post-processing is a helpful method in controlling the fabrication cost also along with increasing the efficiency of production (Espera et al., 2019). Fused deposition modeling (FDM) is an AM technique that utilizes the material in the form of feedstock filament and deals in the fabrication of conceptual parts (Jayanth et al., 2018). It mainly utilizes thermoplastic polymers and their composites such as PLA, acrylonitrile butadiene styrene (ABS), polyether ether ketone (PEEK), polycaprolactone (PCL), thermoplastic polyurethane (TPU), etc. (Prakash et al., 2021) Recent developments in 3D printing materials ensures the printability of biodegradable and biocompatible sensing (wearable or health monitoring sensors) devices (Ling et al., 2020, Guo et al., 2021). PLA-composite-based electronic components are found to help develop a sensor/antenna to analyze the biomedical behavior of the material. PLA reinforced with a measurable quantity of hydroxyapatite (HAP), and chitosan (CS) is found to be a desirable composite for biomedical or tissue engineering applications (Ranjan et al., 2020). HAP and CS replicate the chemical structure of bones and teeth and are found to be suitable for the regeneration of bones (Vaz et al., 1999, Zhu et al., 2002, Suyatma et al., 2004, Shen et al., 2000). Implantable medical sensors/antenna was reported as the latest development for telemetry in biomedical applications (Harysandi et al., 2020). For miniaturization along with improved performance, patch design (in microstrip antennas) attracts a lot of researchers because of their design and conformability (Soontornpipit et al., 2004, Kiziltas et al., 2003, Zhou et al., 2007). Maintaining biocompatibility for a long period, and preventing an antenna from the short-circuit (if the metallic part of the antenna comes in direct contact with human tissues) are the important issues that need to be resolved (Kiourti and Nikita, 2012). Covering the metallic patch with a superstrate layer (dielectric material) prevents the device from a short-circuit and maintains biocompatibility for long-term implantable devices (Karacolak et al., 2008). The literature review reveals that the development of antennas/sensing actuators was commonly reported for implantable devices but hitherto, little has been identified about the prototype development of bio-sensor and antennas to be used as an implant itself utilizing recycled material (from 3D printing waste) in a lab-scale environment.

To ascertain the research gap, a detailed bibliographic analysis has been performed based on the database extracted from the "web of science" platform for the past 15 years. The keywords: sensor, environment, antenna, and economy were searched combinedly on the web of science database platform and the refined (web of science indexed) literature was obtained for the research that happened in the past 1.5 decades. The research data from the existing literature was obtained in the form of a plain text file and gets imported into the VOS viewer software (bibliographic analysis tool). At the beginning of the analysis, the minimum no. of occurrences of a term was selected as 3, based upon that total of 3994 terms extracted from the literature. Out of these terms, 203 meet the threshold value and based upon this, the networking diagram (Figure 5.1) has been constructed which contains 5 clusters that represent the major research reported in a particular area of application.

Table 5.1 contains the information of the terms extracted from the literature obtained through the web of science database. The no. of occurrence of a term and based on that the relevance score for a particular term was calculated in the VOS viewer software itself. Based on the relevance score of the terms, 60% most relevant terms contribute to the construction of a bibliographic diagram.

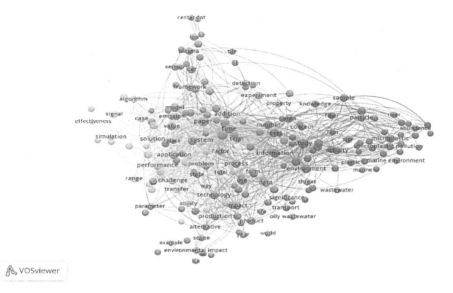

FIGURE 5.1   Bibliographic diagram based on keywords- sensor, antenna, environment, and economy.

TABLE 5.1

Relevance Score and No. of Occurrences of the Terms Used to Draw the Network Diagrams

S.No.	Term	Occurrences	Relevance Score
1	Ability	5	0.5293
2	Abundance	6	3.1856
3	Account	5	0.5714
4	Activity	10	0.7073
5	Addition	16	0.3371
6	Air	4	0.6033
7	Algorithm	10	1.1818
8	Alternative	7	0.5685
9	Amount	7	0.1773
10	Analysis	25	0.1654
11	Animal	4	0.4982
12	Antenna	7	1.4632
13	Application	21	0.3615
14	Approach	20	0.3995
15	Aquatic ecosystem	3	2.4001
16	Area	14	0.2036
17	Article	6	0.2775
18	Atmosphere	3	0.4513
19	Basis	4	1.0535

(Continued)

TABLE 5.1 *(Continued)*
Relevance Score and No. of Occurrences of the Terms Used to Draw the Network Diagrams

S.No.	Term	Occurrences	Relevance Score
20	Calibration	4	0.8576
21	Carbon	4	1.118
22	Case	14	0.8438
23	Case study	3	1.3156
24	Center dot	3	5.9141
25	Challenge	12	0.5556
26	Characteristic	8	0.2751
27	China	6	0.5946
28	Combination	7	0.6798
29	Comparison	7	0.2426
30	Composition	6	2.4892
31	Concentration	4	1.1767
32	Concern	9	0.516
33	Connection	3	1.0762
34	Consideration	4	0.6969
35	Control	9	0.3109
36	Country	4	0.3686
37	Current density	3	1.0638
38	Data	11	0.1491
39	Degrees c	4	0.8775
40	Deployment	3	2.2493
41	Detection	8	0.8748
42	Development	11	0.4331
43	Device	9	0.6366
44	Difference	8	0.3137
45	Discharge	7	0.688
46	Distribution	12	1.5254
47	Economy	5	0.2966
48	Effect	21	0.2091
49	Effectiveness	3	1.7256
50	Efficiency	14	0.6425
51	Emission	8	0.7551
52	End	8	0.2845
53	Energy	13	0.3564
54	Environment	21	0.2771
55	Environmental impact	5	0.9345
56	Environmental threat	3	0.6708
57	Equation	8	0.9714
58	Equilibrium	4	1.3679
59	Estimation	3	0.6495
60	Example	3	0.9283
61	Experiment	10	0.2811

(Continued)

TABLE 5.1 (*Continued*)
Relevance Score and No. of Occurrences of the Terms Used to Draw the
Network Diagrams

S.No.	Term	Occurrences	Relevance Score
62	Experimental result	3	1.113
63	Factor	9	0.311
64	Fate	4	1.9924
65	Feasibility	3	1.7659
66	Fiber	6	1.3971
67	Field	9	0.8624
68	Film	3	1.3892
69	First example	3	5.4579
70	First wall	3	2.6464
71	Form	6	0.2714
72	Formation	8	0.4411
73	Fragment	3	3.0765
74	Framework	13	1.2831
75	Frequency	6	1.251
76	Generator	13	0.5647
77	GHz	4	2.4091
78	Hand	5	0.6771
79	Human health	5	0.7846
80	Iii	4	1.557
81	Impact	13	0.27
82	Impact category	4	1.081
83	Importance	4	0.5223
84	Improvement	3	1.2909
85	Influence	8	0.6288
86	Information	20	0.0764
87	Integration	4	1.0643
88	Interest	6	0.6128
89	Investigation	9	0.3229
90	Ion	8	2.3889
91	Iter	11	1.3994
92	Jet	6	1.7865
93	Knowledge	7	0.7712
94	Lack	4	1.4086
95	Land	3	0.6454
96	LCA	3	1.4434
97	Level	20	0.1553
98	Life	6	0.2881
99	Light	3	0.7146
100	Load	6	0.5908
101	Location	4	0.9296
102	Magnetic field	5	0.6546
103	Marine	3	1.1652

(*Continued*)

TABLE 5.1 (*Continued*)
Relevance Score and No. of Occurrences of the Terms Used to Draw the Network Diagrams

S.No.	Term	Occurrences	Relevance Score
104	Marine environment	5	1.7088
105	Marine organism	4	1.172
106	Mechanism	11	1.0705
107	Methodology	3	0.5623
108	Microplastic	16	1.9368
109	Microplastic abundance	3	2.7362
110	Microplastic contamination	5	2.656
111	Microplastic pollution	11	2.0496
112	Model	21	0.5354
113	MOF	8	3.081
114	Mussel	3	2.9182
115	Nature	3	0.9794
116	Need	7	0.3392
117	Number	12	0.1913
118	Occurrence	5	1.824
119	Oily wastewater	4	0.6991
120	Order	14	0.2928
121	Outlook	3	1.0158
122	Paper	34	0.3175
123	Parameter	7	0.9081
124	Particle	12	1.3894
125	Performance	27	0.4657
126	Perspective	9	0.2846
127	Place	3	0.9644
128	Plasma	9	1.8288
129	Plastic	7	1.2128
130	Plastic pollution	4	1.8947
131	Pollutant	4	1.5812
132	Polyester	3	3.9989
133	Polyethylene	4	3.7092
134	Polypropylene	4	3.7092
135	Potential	8	0.491
136	Power	10	0.735
137	Power distribution	7	0.5636
138	Prediction	3	0.4185
139	Presence	5	0.2909
140	Present study	4	0.3855
141	Problem	14	0.1971
142	Process	18	0.1117
143	Product	13	0.2765
144	Production	10	0.533
145	Property	9	0.4145

(*Continued*)

TABLE 5.1 (*Continued*)
Relevance Score and No. of Occurrences of the Terms Used to Draw the Network Diagrams

S.No.	Term	Occurrences	Relevance Score
146	Range	10	0.7946
147	Rate	15	0.2917
148	Reuse	3	0.9052
149	Recent progress	4	0.4664
150	Recommendation	3	1.1801
151	Reduction	3	0.538
152	Research	11	0.6307
153	Respect	4	2.0793
154	Review	11	0.2902
155	Risk assessment	4	0.4601
156	River	4	2.6962
157	Role	9	0.2674
158	Sample	11	1.1744
159	Scale	9	0.2328
160	Scenario	9	0.6918
161	Scope	6	0.8805
162	Sea	6	2.0739
163	Sediment	5	2.3649
164	Sensor	11	1.6018
165	Show	4	0.7662
166	Signal	9	1.1819
167	Significance	5	0.4221
168	Simulation	7	1.7295
169	Simulation result	3	1.8501
170	Size	8	0.862
171	Soil	3	0.6085
172	Solution	10	0.5296
173	Source	17	0.17
174	Spite	3	1.1171
175	State	7	0.6096
176	Strategy	6	0.775
177	Strength	7	0.4218
178	Structure	12	0.6979
179	Study	37	0.1671
180	Surface water	5	2.3276
181	System	39	0.1431
182	Technology	18	0.2735
183	Term	7	0.3685
184	Threat	7	0.5312
185	Tile	4	1.524
186	Time	29	0.0819
187	Total	3	0.781

(*Continued*)

TABLE 5.1 (*Continued*)

Relevance Score and No. of Occurrences of the Terms Used to Draw the Network Diagrams

S.No.	Term	Occurrences	Relevance Score
188	Transfer	10	0.3695
189	Transport	7	0.2832
190	Treatment	12	0.3043
191	Type	14	0.4491
192	Understanding	6	0.7941
193	Use	16	0.108
194	Value	12	0.7216
195	Vector	5	1.3425
196	View	3	0.9794
197	Wall	4	1.8255
198	Wastewater	4	0.6721
199	Water	14	0.3441
200	Way	6	0.2607
201	Work	9	0.6807
202	World	3	0.5765
203	Year	7	0.4891

Further, to perform the gap analysis, the nodes (economy; environment; antenna; and sensor) have been highlighted (Figure 5.2 a–d) in the main networking diagram (Figure 5.1). Figure 5.2 (a) shows that certain technological advancements have been done to make the end product economical but less has been reported on the reuse of bio-plastics (procured from natural sources) to develop a sensor or an antenna,

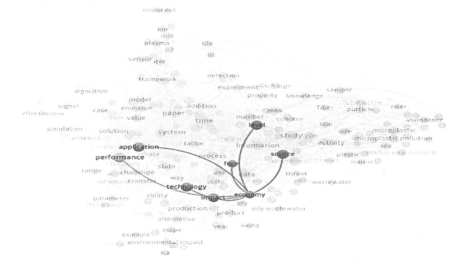

FIGURE 5.2   Research gap analysis by highlighting the nodes: (a) economy. (*Continued*)

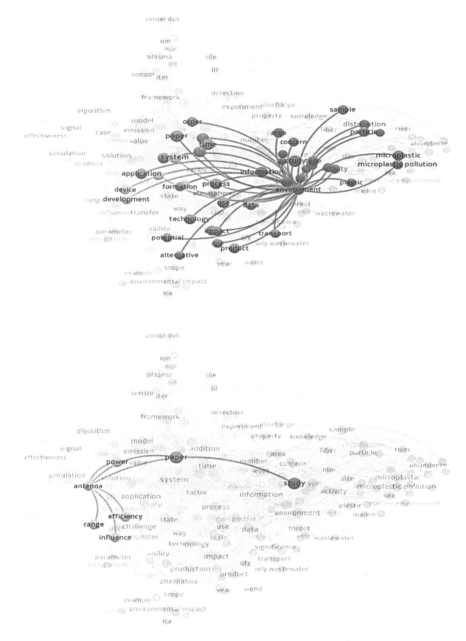

FIGURE 5.2 (*Continued*) Research gap analysis by highlighting the nodes: (b) environment; (c) antenna.

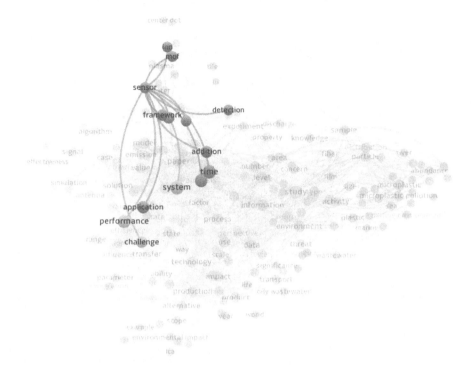

FIGURE 5.2 (*Continued*)   Research gap analysis by highlighting the nodes: (d) sensor.

similarly Figure 5.2 (b) highlights that certain studies have been reported on the impact of microplastics on the environment and its solutions but hitherto limited has been identified on the use of recycled material (3D printing waste) for product development to minimize the solid plastic waste and the emission of harmful gases from such wastage. Moving ahead, Figure 5.2(c) represents the certain parameters that influence the performance of an antenna when it was used in the form of an implantable device but very less has been identified on the use of lab-scale techniques used to improve the efficiency of an antenna. Finally, Figure 5.2(d) represents the application of environmental sensors for the detection of impact on surroundings but very less has been noted on the use of bio-sensors for health-monitoring of patients and the solutions needed to improve the performance by creating the simulated environment inside the lab.

5.2 METHODOLOGY

In this study, a biodegradable and biocompatible material has been used to prepare the composite (PLA-Hap-CS) that was used to fabricate a biosensor. After preparing the composition, the rheological properties of the polymer-based composite were analyzed as per ASTM D1238 by performing the melt flow index (MFI) (Teresa et al., 2007). Further to prepare the feedstock filament, the composition was processed through TSE (at T=180; N=85 rpm, and m=15 kg) for uniform blending of the reinforcements. Following this, to check the recyclability of the material in multiple extrusion cycles, the prepared filaments were shredded uniformly using a mechanical shredder and further extruded through SSE in the form of filament with a uniform diameter (1.75 mm). To determine the mechanical properties of the composite, tensile testing of prepared filaments was ascertained on the universal tensile testing machine (UTM). Going on, the 3D printing of the bio-sensor was carried out on a specific parametric setting (layer height=0.2 mm, heat end temperature=210°C, infill density=100%). Dielectric properties of the substrate based upon ring resonator test were carried out on a VNA. Moving ahead, the hydro-thermal actuation was performed using the substrate and shows some mechanical displacement of the actuator when a hydro-thermal stimulus was provided to it. Based upon the calculated dielectric properties, the designing of the patch antenna has been accomplished in the HFSS software, and to determine the scattering parameters of the antenna, the designed antenna is then simulated in HFSS software itself to identify the radiation behavior of the antenna. Based on simulated results, it was decided whether to fabricate the antenna or not. The detailed methodology is shown in Figure 5.3.

5.3 EXPERIMENTATION

5.3.1 RHEOLOGICAL INVESTIGATION

The rheological characteristics of the prepared composition (PLA-90%; Hap-8%; CS-2%) as per the existing literature have been investigated by performing MFI according to the ASTM D-1238 standard (commonly applied method for thermoplastic polymers). The material was processed through a small die having dimensions 2.095 x 8 mm. The temperature of the heater was taken as 180°C, and the load applied was 2.160 kg, the material was extruded continuously at a specific rate and observed for a period of 10 min.

5.3.2 MECHANICAL PROPERTIES

The tensile testing of the prepared filaments with multi-stage recycling was ascertained on UTM, the prepared samples were having 100 mm length (overall) with 50 mm gauge length, and 25 mm grip separation length. The filament samples were clamped in between the fixtures at both ends and load was applied at a rate of 30 mm/min. After being deformed plastically, the samples were broken at a point, and the mechanical properties (peak load; peak stress; peak strain; young's modulus (E); and modulus of toughness (MOT) were recorded successfully.

5.3.3 MATERIAL EXTRUSION AND 3D PRINTING

The material extrusion of the PLA-Hap-CS-based composite has been carried out initially on TSE for uniform blending by considering the input parameters T=180°C, N=85 rpm, and m=15 kg. For ascertaining the recycling of the composite, the prepared feedstock filaments were being shredded mechanically and the crushed material was then processed through SSE for another extrusion cycle by varying the input

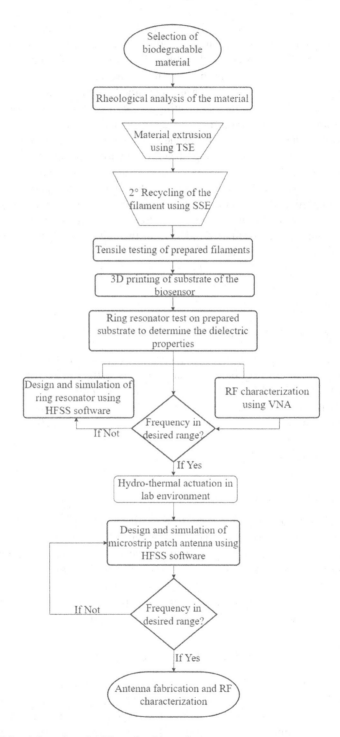

FIGURE 5.3   Adopted methodology for this study.

parameters T=180°C, N=6 rpm, and n=3 shredding cycle. Before 3D printing, the CAD file of parts was generated in Solidworks software and later converted into STL format. The conductive parts (i.e., the ring, feed lines, and ground plane) were fabricated by direct metal laser sintering (DMLS) whereas, the substrate was prepared on an open-source 3D printer (Creality- Ender 3 pro). The following parametric settings were chosen while 3D printing of parts- 0.03 mm layer thickness, 100% infill density for DMLS setup, whereas for polymer printer it was- 0.2 mm layer thickness, 210°C heat end temp., 100% infill density, 80°C bed temperature, and 45° raster angle.

5.3.4 DESIGN AND SIMULATION OF BIO-SENSOR

The dielectric properties of the substrate have been carried out by performing a ring resonator test to design a bio-sensor (based on the principle of microstrip antenna). It is selected as one of the methods to ascertain the dielectric properties of the material because of its highlighting features namely; compact in size, easy to fabricate, and high-quality factor with minimum radiation losses (Ahmed et al., 2017). This method comprises a conducting ring and two feed/transmission lines made up of conducting material separated by a small gap from the ring known as a coupling gap (0.1 to 1 times the width of the feed lines) used to calculate the ε_r and tan δ of the material. The design and simulation of the sensor have been performed using HFSS software (Figure 5.4). The operating frequency range was selected as 1 to 4 GHz while performing the simulation and output will appear in the form of insertion loss vs frequency. The simulated results show that the ring resonator resonated at 2.50 GHz and the S_{21} value was -35.39 dB.

5.3.5 HYDRO-THERMAL ACTUATION

In this study, the actuation of the sample has been carried out in the lab environment. Initially, the sample was mechanically deformed using some mechanical loading, after which the stimulus has been provided to the substrate in the form of water

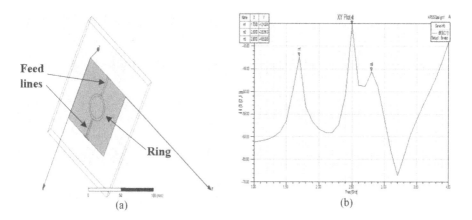

FIGURE 5.4 (a) 3D model of biosensor designed in HFSS software, (b) S_{21} vs RF.

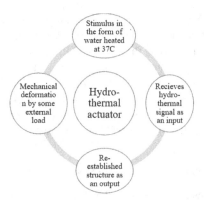

FIGURE 5.5 Control loop actuation process.

heated at 37°C (i.e., the human body temperature) as an input signal. Following this, the actuator performs the task in the form of some mechanical movement to regain the deformed shape of the substrate (Figure 5.5).

5.3.6 MICROSTRIP PATCH ANTENNA

The microstrip patch antenna (MPA) has been designed using HFSS software and based upon that the simulated RF results were determined. The dimensions of the patch were calculated based upon certain input parameters i.e., resonant frequency(f_r)=2.45 GHz; the height of the substrate(h)=1 mm, characteristic impedance (Z_0) =50 Ω; and ε_r=3.32 (obtained from ring resonator test). While designing the antenna (Figure 5.6), the transmission lines were selected to be a microstrip line feed due to their ease of fabrication, simplicity in modeling, and being helpful in

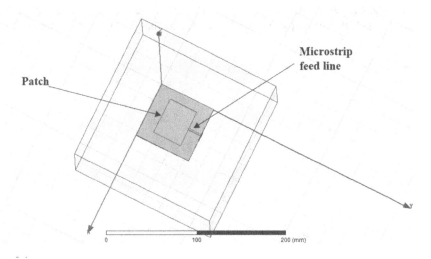

FIGURE 5.6 Designed antenna in HFSS software.

impedance matching also. The following equations were useful while designing an antenna (Manab et al., 2018): -

The width (W) of the patch is given by;

$$W = \frac{C}{2f_r} \sqrt{2/(\varepsilon_r + 1)} \tag{5.1}$$

To determine the fringe factor (ΔL), the effective dielectric constant (ε_{eff}) needs to be calculated i.e.,

$$\varepsilon_{eff} = \left[\frac{(\varepsilon_r + 1)}{2} \right] + \left[\left(\frac{(\varepsilon_r - 1)}{2} \right) \left(1 + 12 \frac{h}{w} \right)^{-0.5} \right] \tag{5.2}$$

$$\Delta L = 0.42h \left[\frac{(\varepsilon_{eff} + 0.3)\left(\frac{w}{h} + 0.264 \right)}{(\varepsilon_{eff} - 0.258)\left(\frac{w}{h} + 0.8 \right)} \right] \tag{5.3}$$

The length (L) of the patch is given by;

$$L = \frac{1}{2f_r \sqrt{(\varepsilon_{eff} \varepsilon_o \mu_o)}} - 2\Delta L \tag{5.4}$$

The dimensions of substrate were given by Equations (5.5) and (5.6) (Bankey and Kumar, 2015);

$$W_s = W + 6h \tag{5.5}$$

$$L_s = L + 6h \tag{6.6}$$

5.4 RESULTS AND DISCUSSIONS

5.4.1 FLOW CHARACTERISTICS

The rheological analysis showcases that the MFI value of PLA-based composite was found to be 11.12 g/10min, and 11.29 g/10min respectively for the multiple stages of 2° recycling. The higher value of MFI indicates a lower value of viscosity (i.e., MFI$\propto$1/viscosity) (Boparai and Singh, 2022). The MFI trend represents that with the successive recycling of the material, the flowability of the material increases which is desirable for developing thinner substrates of the sensors and antennas.

5.4.2 TENSILE BEHAVIOR

The extrusion of feedstock filaments in this study has been done based on the design of the experiment (DOE) by varying input parameters (shredding cycles; Temperature of the hot end; Rotation speed of screw) to achieve better mechanical properties.

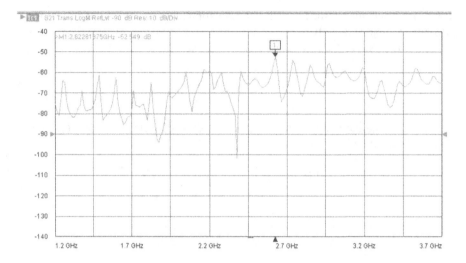

FIGURE 5.7 S_{21} vs operating frequency for ring resonator test.

UTM results represent that, the filament sample prepared with 9- shredding cycles, 170°C- screw temperature, and rotation speed-6 rpm indicates the best mechanical properties (i.e., peak strength: 45.45 MPa, Young's modulus: 1803.57 MPa, MOT: 0.86 MPa, and peak strain: 0.025). Since the composite was prepared for biomedical applications (orthopedic implant), therefore the peak strength along with Young's modulus is one of the desired properties for further analysis.

5.4.3   RF CHARACTERISTICS OF BIOSENSOR

The scattering parameters for the designed ring resonator were obtained through VNA (Figure 5.7), which consists of two ports for measuring the response of the sensor. Before experimenting, the calibration of the VNA at both the ports has been done by using some standards (i.e., load; open, and short) using the calibration tool, and the impedance load was kept at 50 Ω (Singh et al., 2022). The S_{21} parameters have been calculated for the frequency range 1 GHz to 4 GHz. The observations from the test show that the ring resonator resonates at 2.62 GHz and the measured insertion loss corresponding to the resonant frequency was -52.549 dB as shown in Table 5.2. The obtained results highlight that the designed ring resonator gives output comparatively to the simulated results. The ε_r and tan δ of the material have been

TABLE 5.2
Calculated Dielectric Properties of the Substrate Material

Resonant Frequency	Insertion Loss	Dielectric Constant	Loss Tangent
2.62	-52.549	3.32	0.0042

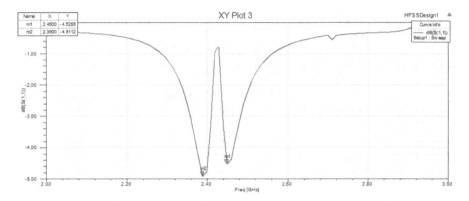

FIGURE 5.8   Simulation results for MPA (S_{11} vs frequency).

calculated by following equations based on existing literature on ring resonators, and the obtained values of ε_r and tan δ are 3.32 and 0.0042 respectively.

5.4.4   ACTUATOR BEHAVIOR

The actuation process starts once the sample receives the input signal (hydrothermal). The deformation produced in the sample by applying external load was 2.98 mm whereas the final length of the sample was found to be 93.05 mm from 90.07 mm. After receiving the hydro-thermal stimulus, the sample recovered the deformation produced by the external load and the final length was found to be 90.26 mm.

5.4.5   MICROSTRIP ANTENNAS

The MPA was designed as per the patch dimensions (L=33.35 mm; W=41.66 mm) calculated by Equations (5.1–5.6) using HFSS software. The designed antenna has been simulated in HFSS software by considering resonant frequency as 2.45 GHz. The antenna performance was analyzed by adding the frequency sweep setup with the frequency range 2–3 GHz and a step size of 0.1. Several peaks were observed in the response diagram between a return loss (S_{11}) and frequency and it was noticed that the antenna resonates at 2.39 and 2.45 GHz in the ISM band with S_{11}= −4.91 and 4.52 dB respectively (Figure 5.8). From the above observations, it may be concluded that the designed antenna resonates in the ISM band and was suitable for fabrication for further analysis.

5.5   ENVIRONMENT AND ECONOMIC JUSTIFICATION

From the environmental and economic point of view, the composite used in this study was biodegradable as well as biocompatible material and hence decomposes with time. Here in this work, the 3D printing-based bio-waste has been utilized for further development of working prototypes or products by performing 2° recycling in multiple stages. Table 5.3 indicates the effect of multi-stage recycling on different

TABLE 5.3

Behavior of Different Properties after Multi-Stage Recycling of the Composite

Property	1st Stage of 2° Recycling	2nd Stage of 2° Recycling
Flow property (MFI in g/10min)	11.12	11.29
MOT (MPa)	1.17	0.79
Peak strength (MPa)	37.21	29.28
Peak strain	0.055	0.051
Young's Modulus (MPa)	685.52	581.96

properties of the prepared composite, the MFI value was increased from 11.12 to 11.29 which represents the recycled composite can be used to prepare lightweight components as the molecular weight of the material decreased with multi-stage of recycling (Spear et al., 2015). While the mechanical properties of the material degrade with the increase in no. of stages of recycling i.e., MOT value decreases by 32.47% after 2nd stage which results in the brittle failure of the components (Kumar et al., 2022), Peak strength decreased by 21.31%, similarly, Young's modulus also reduced by 15.10%, whereas the peak strain reduces by 7.27%. As observed from Table 5.3, the mechanical properties diminish with the increase in the stage of recycling but a little impact was observed on the peak strain of the composite which represents the composite was suitable for preparing components where higher elasticity was desired.

5.6   CONCLUSIONS

The following conclusions were drawn from this work:

- In this study, the secondary recycling (in multiple stages) of the PLA-based material was ascertained and it was found that very less degradation in the mechanical properties of the material up to two heat extrusion cycles, which may be helpful in the waste management of bio-based polymers.
- As regards the economic aspect, the material is suitable for the development of bio-sensors by utilizing recycled material which makes it a cost-effective method to produce 3D printed bio-sensors.
- Simulation results of the antenna represent that it was suitable for fabrication and may be utilized as an implantable antenna as it resonates in the ISM band and can transmit a signal which was remotely monitored.
- Actuation results highlight that the actuator was sensitive to the hydrothermal signal and performs the task in the form of some mechanical displacement.
- The RF behavior of biosensors along with the mechanical properties of the material represents that the device can efficiently be used for biomedical telemetry/biomedical applications (specifically for implants and scaffolds) with high mechanical strength.

REFERENCES

Ahmed M. I., Ahmed M. F., Shaalan A. A. (2017). Investigation and comparison of 2.4 GHz wearable antennas on three textile substrates and their performance characteristics. *Open Journal of Antennas and Propagation*, 5(3):110–120.

Bankey V. and Kumar N. A. (2015). Design and performance issues of microstrip antennas. *Int J Sci Eng*, 6: 1572–1580.

Boparai K. S., Singh R. (2022). Hydrothermal stimulus for 4D capabilities of PA6-Al-Al2O3 composite. *4D Printing*, pp. 121–145.

Espera A. H., Dizon J. R., Chen Q., Advincula R. C. (2019). 3D printing and advanced manufacturing for electronics. *Progress in Additive Manufacturing*, 4(3):245–267.

Guo Y., Chen S., Sun L., Yang L., Zhang L., Lou J., You Z. (2021). Degradable and Fully Recyclable Dynamic Thermoset Elastomer for 3D-Printed Wearable Electronics. *Advanced Functional Materials*, 31(9):2009799.

Harysandi D. K., Oktaviani R., Meylani L., Vonnisa M., Hashiguchi H., Shimomai T., Luini L., Nugroho S., Aris N. A. (2020). *International Telecommunication Union-Radiocommunication Sector (ITU-R)*, Performance to estimate Indonesian rainfall. *Telkomnika*, 1;18(5):2292–2303.

Jayanth N., Senthil P., Prakash C. (2018). Effect of chemical treatment on tensile strength and surface roughness of 3D-printed ABS using the FDM process. *Virtual and Physical Prototyping*, 3;13(3):155–163.

Karacolak, T., Hood, A. Z., & Topsakal, E. (2008). "Design of a Dual-Band Implantable Antenna and Development of Skin Mimicking Gels for Continuous Glucose Monitoring," *IEEE Transactions on Microwave Theory and Techniques*, pp. 1001–1008.

Kiourti A., Nikita K. S. (2012). A review of implantable patch antennas for biomedical telemetry: Challenges and solutions [wireless corner]. *IEEE Antennas and Propagation Magazine*, 31;54(3):210–228.

Kiziltas G., Psychoudakis D., Volakis J. L., Kikuchi N. (2003). Topology design optimization of dielectric substrates for bandwidth improvement of a patch antenna. *IEEE Transactions on Antennas and Propagation*, 14;51(10):2732–2743.

Kumar S., Singh R., Singh T. P., Batish A., Singh M. (2022). Multi-stage Primary and Secondary Recycled PLA Composite Matrix for 3D Printing Applications. *Proceedings of the National Academy of Sciences, India Section A: Physical Sciences*, 1–22.

Ling H., Chen R., Huang Q., Shen F., Wang Y., Wang X. (2020). Transparent, flexible, and recyclable nano paper-based touch sensors fabricated via inkjet printing. *Green Chemistry*, 22(10):3208–3215.

Maddipatla D., Narakathu B. B., Atashbar M. (2020). Recent progress in manufacturing techniques of printed and flexible sensors: a review. *Biosensors*, 10(12):199.

Manab N. H., Baharudin E., Seman F. C., Ismail A. (2018). 2.45 GHz Patch antenna based on thermoplastic polymer substrates. *IEEE International RF and Microwave Conference*, pp. 93–96.

Prakash C., Senthil P., Sathies T. (2021). Fused deposition modeling fabricated PLA dielectric substrate for microstrip patch antenna. *Materials Today: Proceedings*, 1;39:533–537.

Ranjan N., Singh R., Ahuja I. P. (2020). Development of PLA-HAp-CS-based biocompatible functional prototype: a case study. *Journal of Thermoplastic Composite Materials*, 33(3):305–323.

Shen F., Cui Y. L., Yang L. F., Yao K. D., Dong X. H., Jia W. Y., Shi H. D. (2000). A study on the fabrication of porous chitosan/gelatin network scaffold for tissue engineering. *Polymer International*, 49(12):1596–1599.

Singh R., Kumar S., Singh A. P., Wei Y. (2022). On comparison of recycled LDPE and LDPE–bakelite composite based 3D-printed patch antenna. *Proceedings of the Institution of Mechanical Engineers, Part L: Journal of Materials: Design and Applications,* 236(4):842–856.

Soontornpipit P., Furse C. M., Chung Y. C. (2004). Design of implantable microstrip antenna for communication with medical implants. *IEEE Transactions on Microwave theory and techniques,* 2;52(8):1944–1951.

Spear M. J., Eder A., Carus M. (2015). Wood polymer composites. *Wood composites,* pp. 195–249.

Suyatma N. E., Copinet A., Tighzert L., Coma V. (2004). Mechanical and barrier properties of biodegradable films made from chitosan and poly (lactic acid) blends. *Journal of Polymers and the Environment,* 12(1):1–6.

Teresa Rodríguez-Hernández M., Angulo-Sánchez J. L., Pérez-Chantaco A. (2007). Determination of the molecular characteristics of commercial polyethylenes with different architectures and the relation with the melt flow index. *Journal of applied polymer science,* 104(3):1572–1578.

Vaz L., Lopes A. B,, Almeida M. (1999). Porosity control of hydroxyapatite implants. *Journal of Materials Science: Materials in Medicine,* 10(4):239–242.

Wiklund J., Karakoç A., Palko T., Yiğitler H., Ruttik K., Jäntti R., Paltakari J. (2021). A review on printed electronics: Fabrication methods, inks, substrates, applications, and environmental impacts. *Journal of Manufacturing and Materials Processing,* 5(3):89.

Zhou, Y., Chen, C. C., Volakis, J. L. (2007). "Dual-Band ProximityFed Stacked Patch Antenna for Tri-Band GPS Applications," *IEEE Transactions on Antennas and Propagation,* pp. 220–223.

Zhu A., Zhang M., Wu J., Shen J. (2002). Covalent immobilization of chitosan/heparin complex with a photosensitive hetero-bifunctional crosslinking reagent on PLA surface. *Biomaterials,* 23(23):4657–4665.

6 Flexible and Wearable Patch Antenna Using Additive Manufacturing for Modern Wireless Applications

Sanjeev Kumar[1], Rupinder Singh[2], Amrinder Pal Singh[1], and Yang Wei[3]
[1]University Institute of Engineering and Technology, Panjab University, Chandigarh, India
[2]National Institute of Technical Teachers Training and Research, Chandigarh, India
[3]Nottingham Trent University, Nottingham, U.K.

CONTENTS

6.1 INTRODUCTION

In the past decade, there is a fast-growing interest in research of wearable antennas integrated into garments for the complete wearable system. Flexible and wearable antennas have attracted researchers due to their application in BAN for reliable and robust connectivity between bodies providing massive IoT (Njogu P. et al., 2020). Wearable antennas are mounted on the body or connected to wearable components such as clothes, caps, etc. providing continuous communication for the parameters to be monitored such as the temperature of the body, pulse rate, heartbeat, location, etc. (Scarpello et al., 2012). Usually, wearable antennas are flexible to provide conformability and easy application on wearable clothes such as fabrics. Wearable antennas are generally referred to as smart wearables providing wireless connectivity.

DOI: 10.1201/9781003194224-6

The mechanical properties of the flexible antennas, including to be bent, wrinkled and stressed, has extended their applications in many fields of modern electronic applications (Lim et al., 2019).

An antenna, (the main part of the wearable antennas), can be a transmitter or receiver based on the application to transmit or receive the signals within a specific range. Microstrip patch antenna (MPA) has a small size, low cost, lightweight, and simple structure making it most suitable for antenna applications. MPA has three sub-parts namely patch or radiating part, ground, and substrate. Patch and ground are conductive materials whereas substrate is a dielectric material. A patch is printed on top of the substrate and the ground is printed on the bottom. There are different methods to excite the antenna (Singh et al., 2022).

Additive manufacturing (AM), often known as 3D printing, is a group of new technologies that allow the production of items by layering materials from the bottom to up. The designs for the AM are prepared by using computer-aided design (CAD), which are to be further converted into a physical model. AM generally reduces the waste because it does not remove any material from the workpiece and simultaneously provides fast production of complex shapes (Xin & Liang, 2017). AM or 3D printing has shown its application in the field of wireless electronics providing the substrates of customized structures (different infill densities, complex shapes, etc.) for the MPA (Mirzaee & Noghanian, 2016). Polymer materials such as acrylonitrile-butadiene-styrene (ABS) and polylactic acid (PLA) have been used widely for printing substrates using 3D printing (Hoyack et al., 2016). Researchers have used AM for printing all parts of the antenna including conductive (Garcia et al., 2013) and dielectric parts (Ren & Yin et al., 2018) or both (McKerricher et al., 2016). Using AM for printing antennas provides an interdisciplinary field comprised of material science and electronics (Tan et al., 2019). AM has also been used to print the antenna on textiles to make smart wearables (Jun et al., 2018). Smart wearables and textiles have been used as substrates in some previous studies (Nacer et al., 2012). Direct printing on textiles by using inkjet printing has been outlined by some previous researchers as a fruitful technique (Jun et al., 2018). Wearable antennas have been printed on leather in the past for sensing applications (Shamim, 2013). Researchers have also investigated armbands (Abbas et al., 2014), watches (Shafqat A. et al., 2018), and glasses as wearable antennas.

The wearable antennas should be flexible to provide comfort to the user. Although ABS and PLA have been widely used for the MPA applications, their nonflexible and nonstretchable nature has eliminated them from wearable applications (Rizwan et al., 2017). Researchers have used NinjaFlex on textiles (using FFF) for wearable antennas (Moscato Stefano et al., 2016). It has been reported that NinjaFlex is a strong and flexible material composed of thermoplastics and rubber (Bahr Ryan et al., 2015). TPU is another alternative for flexible and wearable antennas (He Xifu et al., 2020). TPU behaves like an elastomer with good recyclability, high flexibility, stretchability, and high mechanical strength (Tian Ming et al., 2014). TPU is chemically stable and easy to process and cost-efficient (Chen Tianjiao et al., 2021) and has great application in the field of footwear, automotive, sports, hose, pipes cables, etc. (Tayfun & Kanbur, 2016). The excellent electrical properties and shape recoverability (He Xifu et al., 2020) have made TPU an excellent candidate for flexible and wearable antennas. 3D printing of TPU for antennas has been reported by researchers (Mukai et al., 2021) for breast hyperthermia.

In this work, the authors have purposed a TPU + Fabric base MPA for wearable applications using FFF (a technique of AM). TPU has been directly printed on cotton + lycra fabric using FFF and both were combined and used as a substrate for antenna design. Further rheological, mechanical, and RF characteristics were investigated.

6.2 LITERATURE REVIEW

(Hosseini Varkiani & Afsahi, 2019) have developed a compact co-planar waveguide fed square slot antenna fabricated on a cotton layer with a dielectric constant of 1.65 for wearable applications. The developed antenna was also compared with non-woven fabric. Kiani et al. (2021) has purposed a Mercedes-Benz logo wearable antenna with a silver plate at an optimal distance to avoid radiation from the human body. The researcher printed an antenna on Rogers 4003C substrate with a bandwidth from 2.20 to 2.56 GHz, SAR levels were also found in control. Washable antennas were purposed by (Scarpello et al., 2012) integrated into garments and a covering of TPU was done using inkjet printing. Njogu P. et al., 2020 designed a 3D printed wearable antenna in the form of a fingernail using ABS. A MPA was printed using aerosol jet technology. Pei et al., 2015 investigated direct printing of ABS, PLA, and nylon on eight different types of woven and knitted fabrics resulting in PLA as the best of the three polymers used whereas woven cotton fabric has shown better results in terms of warp, print quality. So, in this work woven cotton fabric has been selected, and initially PLA was printed on it. But due to poor printability and less flexibility, TPU was selected for direct printing on fabric.

6.3 RESEARCH GAP

To ascertain the research gap for TPU + textile-based 3D printed wearable antenna, bibliographic analysis was performed by using a web of science database. Initially, the search was performed with a keyword as a wearable antenna, and 1353 results were obtained (for the selected period 2000–2022). Out of these results, the most recent 1000 results have been shortlisted for further analysis using VOS viewer open-source software. By selecting a minimum number of occurrences of the term as '6', out of 16066 terms, 836 met the threshold. For each of 836 term relevance score was calculated and based upon this score, 60% most relevant terms (502) were selected for final analysis. By using these 502 terms, a networking diagram as a bibliographic analysis has been prepared (Figure 6.1).

Similarly, another search was performed for the keyword "3D printed wearable antenna" and 21 results were obtained. Figures 6.2 and 6.3 respectively show bibliographic analysis and gap analysis for the selected keyword.

Finally, for 3D printed wearable antenna + fabric as a keyword, no result was obtained in the web of science core collection. The literature review reveals that significant studies have been reported on wearable antennas. Also some work has been reported on 3D printed wearable antennas. But no work has been reported on based TPU + fabric-based 3D printed substrate for wearable antennas and their

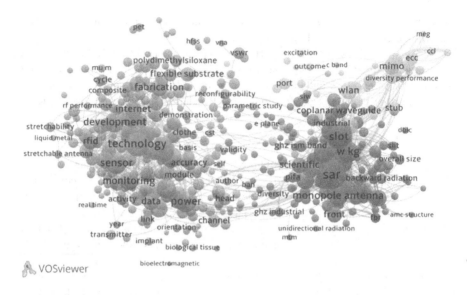

FIGURE 6.1 Bibliographic analysis based upon keyword wearable antenna.

mechanical, rheological and RF characterization for wireless applications in the web of science core collection. Further, the TPU + fabric-based substrates in wearable antenna may be treated as a novel way of secondary recycling of TPU resulting in the development of a high-end value-added product. So, in this study, a concept of TPU + fabric-based wearable antenna has been purposed. The substrate of TPU was directly printed on the woven cotton fabric on FFF for the preparation of a wearable antenna.

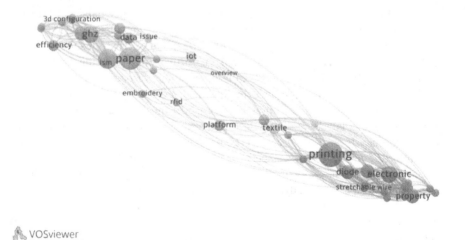

FIGURE 6.2 Bibliographic analysis based upon keyword 3D printed wearable antenna.

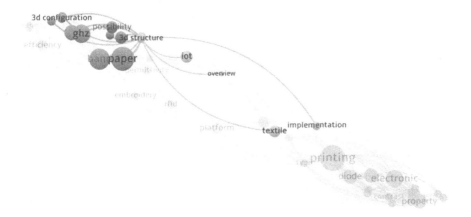

FIGURE 6.3   Gap analysis based upon keyword 3D printed wearable antenna.

6.4   MATERIALS AND METHODS

The methodology used in this work is shown in Figure 6.4.

Cotton + Lycra fabric was selected for the experimentation. To select the polymer for the experimentation, printing of TPU and PLA was done directly on fabric. The TPU showed better printability as compared to the PLA on fabric as shown in Figure 6.5.

As a result, TPU was selected for further experimentation. MFI value of TPU was tested using ASTM D 1238 standards at a temperature of 190°C with a 2.16 kg load applied as shown in Figure 6.6.

Further, the filament of TPU was extruded using a SSE at 170°C and 7 rpm as shown in Figure 6.7.

Dumbbell samples using TPU filament were printed on fabric also using FFF as shown in Figure 6.8 at an infill density of 80%.

After this, the mechanical testing of both samples was performed on universal testing machine (UTM) as shown in Figure 6.9.

The filament was then used to print the substrate of size 60 mm×60 mm×0.65 mm directly on the fabric using FFF as shown in Figure 6.10 using the parameters shown in Table 6.1.

After this, the ring resonator was designed using the equations from previous literature by (Singh et al., 2022).

TABLE 6.1
Printing Parameters

S. No.	Nozzle Temperature (°C)	Bed Temperature (°C)	Printing Speed (mm/sec)	Infill density (%)
1.	220	50	18	80

FIGURE 6.4 Methodology adopted for the present study.

For analysis purposes in this work $\varepsilon_r = 2.5$ has been assumed initially and the ring resonator has been designed for 2.45 GHz.

$$2\pi R = n\lambda_g \ \ for \ n = 1, 2, 3 \tag{6.1}$$

Here, R is the mean radius of the ring and n is the harmonic order of resonance. Guided wavelength (λ_g) was calculated by using the following relation:

$$\lambda_g = \frac{c}{\sqrt{\left(\varepsilon_{eff}\right)}} * \frac{1}{f} \tag{6.2}$$

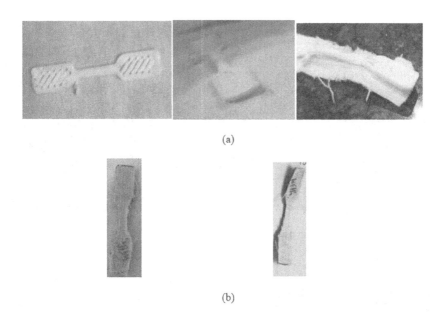

(a)

(b)

FIGURE 6.5 Printing and adhesion (a) PLA on fabric, (b) TPU on fabric.

FIGURE 6.6 MFI tester.

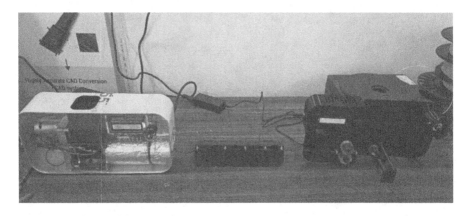

FIGURE 6.7   Wire extrusion using SSE.

(a)   (b)

FIGURE 6.8   Dumbbells printed (a) TPU only, (b) TPU + Fabric.

(a)   (b)

FIGURE 6.9   UTM testing of (a) TPU only, (b) TPU + fabric.

FIGURE 6.10   Printing of substrate on fabric.

c is the velocity of light, f is the resonant frequency, λ_g is the guided wavelength or wavelength in substrate and ε_{eff} is the effective dielectric constant and it is given by the relation of:

$$\epsilon_{eff} = \left[\frac{\epsilon_r + 1}{2}\right] + \left[\left(\frac{\epsilon_r - 1}{2}\right)\left(1 + 12\frac{h}{w}\right)^{-0.5}\right] \tag{6.3}$$

Where, ϵ_r is relative dielectric constant, w is the width of the ring and h is the substrate thickness.

Feed line (transmission line) length was computed using the following relation:

$$\text{Feed line length} = \frac{\lambda_g}{4} \tag{6.4}$$

The inner and outer radius of the ring was computed by subtraction and addition of half of the width of the microstrip to the mean radius respectively. It is given by:-

$$\text{Inner radius: } R - \frac{w}{2} \tag{6.5}$$

$$\text{Outer radius: } R + \frac{w}{2} \tag{6.6}$$

The calculated values of the ring resonator are shown in Table 6.2.

A simulation was done using calculated dimensions and it was observed that the first peak of the ring resonator was at 2.45 GHz defining the accuracy of calculated dimensions as shown in Figure 6.11.

The physical model of the ring resonator was prepared using copper tape (thickness 0.08 mm) like a ring, ground, and feed lines, and SubMiniature version A (SMA) connectors (impedance 50 Ω) were soldered to the feed lines as shown in Figure 6.12.

TABLE 6.2

Calculated Dimensions of the Ring Resonator

Parameters	Values
Mean radius	13.5 mm
Width of the feed line	1.91 mm
Feed line length	10.821 mm
Coupling gap	0.2469 mm
The total length of the substrate	60 mm
Total width of the substrate	54 mm
The inner radius of a ring	12.55 mm
The outer radius of a ring	14.55 mm
Substrate used	TPU + Fabric

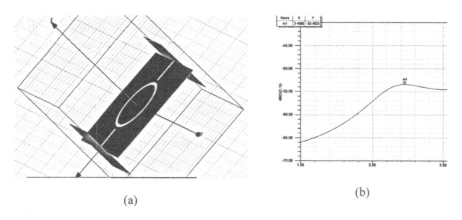

(a)

(b)

FIGURE 6.11   Ring resonator simulation (a) Design, (b) first resonating frequency.

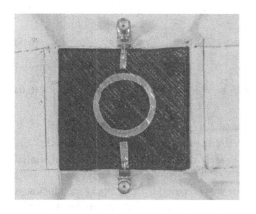
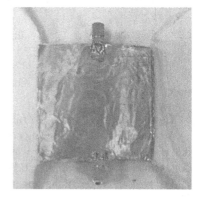

FIGURE 6.12   A ring resonator.

6.3   RESULTS AND DISCUSSION

MFI of TPU was observed as 29.75 g/10min using the ASTM D1238. The tensile testing of TPU suggested that it has a stiffness of 3.897 N/mm and peak stress of 8.275 MPa without breaking due to high elasticity whereas the TPU printed on cloth has a stiffness of 6.658 N/mm and peak stress of 6.6 MPa. The results of TPU on the fabric were observed till the fabric was torn and then the peak starts falling in the stress-strain curve. The stress-strain curve and tensile testing results for both samples are shown in Figure 6.13 and Table 6.3.

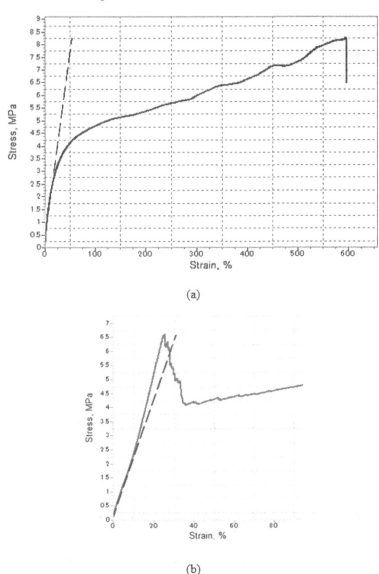

(a)

(b)

FIGURE 6.13   Stress-Strain curves (a) For TPU only, (b) For TPU printed on fabric.

TABLE 6.3
Tensile Testing Results

Sample	Stiffness (N/mm)	Peak Stress (MPa)	Peak Elongation (%)	Young's Modulus (MPa)
TPU	3.897	8.275	595.976	15
TPU on fabric	6.658	6.6	134.532	21

Further RF characteristics of the ring resonator were tested on VNA as shown in Figure 6.14 which gives the results as shown in Table 6.4. It was observed that the ring resonates at 2.80 GHz with S_{21} of -41.8843 dB. Further, the dielectric properties of the combined substrate (TPU + Fabric) can be calculated using the equations given by (Singh et al., 2022).

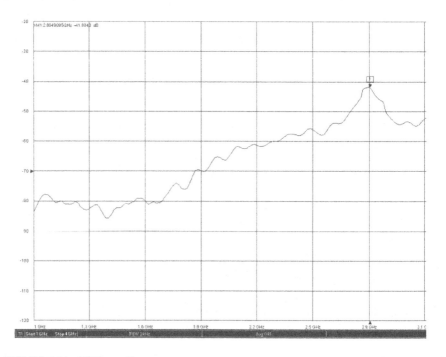

FIGURE 6.14   VNA results.

TABLE 6.4
RF Characteristics of Ring Resonator Which Can Be Used as Wearable Applications

Material	Resonating Frequency (GHz)	Insertion Loss (dB)
TPU + Fabric	2.80	-41.8843

6.4 CONCLUSION

A wearable antenna-based sensor has been successfully prepared. Based on VNA, the RF characteristics of the ring resonator suggested resonating frequency at 2.80 GHz with an S_{21} of 41.8843 dB. Also, in this study, TPU was successfully printed on cotton + lycra fabric using FFF. For the fabrication of wearable antenna-based sensors, the rheological, mechanical, and RF characteristics of the purposed antenna have been investigated. The mechanical testing of TPU and TPU + woven fabric suggested stiffness of 3.897 N/mm and 6.658 N/mm respectively, which may be tuned to get acceptable sensing capabilities.

For future investigations, the effect of strain as the 4D property may be investigated for ascertaining the RF characteristics of the antenna. Also, different thermoplastic materials may be investigated for printing the wearable antennas.

ACKNOWLEDGMENT

The research has been partially funded under NTU-PU collaborated project titled "Wearable 3D printed patch antenna." The most part has been done at the Manufacturing research lab, GNDEC, Ludhiana, and National Institute of Technical Teachers Training and Research, Chandigarh (under PU Chandigarh).

REFERENCES

Bahr, R., Le, T., Tentzeris, M. M., Moscato, S., Pasian, M., Bozzi, M., Perregrini, L. (2015). RF characterization of 3D printed flexible materials-NinjaFlex Filaments. *European Microwave Week 2015: "Freedom Through Microwaves," EuMW 2015 - Conference Proceedings; 2015 45th European Microwave Conference Proceedings, EuMC*, 742–745. https://doi.org/10.1109/EuMC.2015.7345870

Chahat, N., Maxim Zhadobov, L. L. C., and Sauleau, R. R. (2012). Wearable endfire textile antenna for on-body communications at 60 GHz. *IEEE Antennas and Wireless Propagation Letters, 11*(10 m), 799–802. https://doi.org/10.1109/LAWP.2012.2207698

Chen, T., Xie, Y., Wang, Z., Lou, J., Liu, D., Xu, R., Cui, Z., Li, S., Panahi-Sarmad, M., and Xiao, X. (2021). Recent advances of flexible strain sensors based on conductive fillers and thermoplastic polyurethane matrixes. *ACS Applied Polymer Materials, 3*(11), 5317–5338.

Farooqui, M.F. and Shamim, A. (2013). Dual-band inkjet printed bow-tie slot antenna on leather. *2013 7th European Conference on Antennas and Propagation, EuCAP 2013*, 3287–3290.

Garcia, C. R., Rumpf, R. C., Tsang, H. H., and Barton, J. H. (2013). Effects of extreme surface roughness on 3D printed horn antenna. *Electronics Letters, 49*(12), 734–736.

He, X., Zhou, J., Jin, L., Long, X., Wu, H., Xu, L., Gong, Y., and Zhou, W. (2020). Improved dielectric properties of thermoplastic polyurethane elastomer filled with core–shell structured PDA@ TiC particles. *Materials, 13*(15), 3341.

Hong, S., Kang, S.H., Kim, Y., and Jung, C.W. (2016). Transparent and flexible antenna for wearable glasses applications. *IEEE Transactions on Antennas and Propagation, 64*(7), 2797–2804.

Hosseini Varkiani, S. M., & Afsahi, M. (2019). Compact and ultra-wideband CPW-fed square slot antenna for wearable applications. *AEU - International Journal of Electronics and Communications, 106*, 108–115. https://doi.org/10.1016/j.aeue.2019.04.024

Hoyack, M., Bjorgaard, J., Huber, E., Mirzaee, M., & Noghanian, S. (2016). Connector design for 3D printed antennas. *2016 IEEE Antennas and Propagation Society International Symposium, APSURSI 2016 - Proceedings*, 477–478. https://doi.org/10.1109/APS. 2016.7695947

Jun, S. Y., Elibiary, A., Sanz-Izquierdo, B., Winchester, L., Bird, D., & McCleland, A. (2018). 3D printing of conformal antennas for diversity wrist-worn applications. *IEEE Transactions on Components, Packaging and Manufacturing Technology, 8*(12), 2227–2235. https://doi.org/10.1109/TCPMT.2018.2874424

Kiani, S., Rezaei, P., & Fakhr, M. (2021). A CPW-fed wearable antenna at ISM band for biomedical and WBAN applications. *Wireless Networks, 27*(1), 735–745. https://doi. org/10.1007/s11276-020-02490-1

Lim, C., Shin, Y., Jung, J., Kim, J. H., Lee, S., & Kim, D. (2019). *Stretchable conductive nanocomposite based on alginate hydrogel and silver nanowires for wearable electronics.* 031502. https://doi.org/10.1063/1.5063657

McKerricher, G., Titterington, D., & Shamim, A. (2016). A Fully Inkjet-Printed 3D Honeycomb-Inspired Patch Antenna[1] G. McKerricher, D. Titterington, and A. Shamim, "A Fully Inkjet-Printed 3D Honeycomb-Inspired Patch Antenna," *IEEE Antennas Wirel. Propag. Lett.*, vol. 15, pp. 544–547, 2016, DOI: 10.1109/LAWP. *IEEE Antennas and Wireless Propagation Letters, 15*, 544–547. https://doi.org/10.1109/ LAWP.2015.2457492

Mirzaee, M., & Noghanian, S. (2016). Additive manufacturing of a compact 3D dipole antenna using ABS thermoplastic and high-temperature carbon paste. *2016 IEEE Antennas and Propagation Society International Symposium, APSURSI 2016 - Proceedings, 1*, 475–476. https://doi.org/10.1109/APS.2016.7695946

Moscato, S., Bahr, R., Le, T., Pasian, M., Bozzi, M., Perregrini, L., and Tentzeris, M.M. (2016). Infill-dependent 3-D-printed material based on NinjaFlex filament for antenna applications. *IEEE Antennas and Wireless Propagation Letters, 15*, 1506–1509.

Mukai, Y., Li, S., & Suh, M. (2021). 3D-printed thermoplastic polyurethane for wearable breast hyperthermia. *Fashion and Textiles, 8*(1). https://doi.org/10.1186/s40691-021-00248-7

Njogu, P., Sanz-Izquierdo, B., Elibiary, A., Jun, S.Y., Chen, Z., and Bird, D., (2020). 3D printed fingernail antennas for 5G applications. *IEEE Access, 8*, 228711–228719.

Pei, E., Shen, J., & Watling, J. (2015). *Direct 3D printing of polymers onto textiles: experimental studies and applications.* June. https://doi.org/10.1108/RPJ-09-2014-0126

Rizwan, M., Khan, M. W. A., Sydanheimo, L., Virkki, J., & Ukkonen, L. (2017). Flexible and Stretchable Brush-Painted Wearable Antenna on a Three-Dimensional (3D) Printed Substrate. *IEEE Antennas and Wireless Propagation Letters, 16*, 3108–3112. https:// doi.org/10.1109/LAWP.2017.2763743

Scarpello, M. L., Kazani, I., Hertleer, C., Rogier, H., & Vande Ginste, D. (2012). Stability and efficiency of screen-printed wearable and washable antennas. *IEEE Antennas and Wireless Propagation Letters, 11*, 838–841. https://doi.org/10.1109/LAWP.2012.2207941

Shafqat, A., Tahir, F. A., & Cheema, H. M. "A Compact Uniplanar Tri-band Antenna for Wearable Smart Watches," *2018 18th International Symposium on Antenna Technology and Applied Electromagnetics (ANTEM)*, 2018, pp. 1–3, DOI: 10.1109/ ANTEM.2018.8572966.

Singh, R., Kumar, S., Singh, A. P., & Wei, Y. (2022). On comparison of recycled LDPE and LDPE–bakelite composite-based 3D printed patch antenna. *Proceedings of the Institution of Mechanical Engineers, Part L: Journal of Materials: Design and Applications, 236*(4), 842–856. https://doi.org/10.1177/14644207211060465

Syed Muzahir Abbas, Karu P., Esselle, Y. R. (2014). An armband-wearable printed antenna with a full ground plane for body area networks. *IEEE Antennas and Propagation Society, AP-S International Symposium (Digest), 35 mm*, 318–319. https://doi. org/10.1109/APS.2014.6904491

Tan, H. W., An, J., Chua, C. K., & Tran, T. (2019). Metallic Nanoparticle Inks for 3D Printing of Electronics. *Advanced Electronic Materials*, *5*(5). https://doi.org/10.1002/aelm.201800831

Tayfun, U., & Kanbur, Y. (2016). *Mechanical, electrical, and melt flow properties of polyurethane elastomer/surface-modified carbon nanotube composites. September.* https://doi.org/10.1177/0021998316666158

Tian, M., Yao, Y., Liu, S., Yang, D., Zhang, L., Nishi, T., and Ning, N. (2015). Separated-structured all-organic dielectric elastomer with large actuation strain under ultra-low voltage and high mechanical strength. *Journal of Materials Chemistry A*, *3*(4), 1483–1491.

Xin, H., & Liang, M. (2017). 3D-Printed Microwave and THz Devices Using Polymer Jetting Techniques. *Proceedings of the IEEE*, *105*(4), 737–755. https://doi.org/10.1109/jproc.2016.2621118

Yin, J. R. and J. Y. (2018). 3D-PRinted Low-Cost Dielectric-Resonator-Based Ultra-Broadband Microwave Absorber Using Carbon-Loaded Acrylonitrile Butadiene Styrene Polymer. *Materials*, *11*(7), 1249.

7 4D Printed Smart Sensor, Actuators, and Antennas

Vinay Kumar[1,2], Rupinder Singh[3],
Inderpreet Singh Ahuja[4], and Sanjeev Kumar[5]
[1]Guru Nanak dev Engineering College, Ludhiana, India
[2]University Centre for Research and Development,
Chandigarh University, Mohali, India
[3]National Institute of Technical Teachers Training
and Research, Chandigarh, India
[4]Punjabi University, Patiala, India
[5]University Institute of Engineering and Technology,
Panjab University, Chandigarh, India

CONTENTS

7.1   INTRODUCTION

The studies reported in the recent past on advanced sensors and actuators prepared by innovative techniques like ultrasonic additive manufacturing, composite structures, and in-situ polarization have highlighted various engineering applications of sensors like visualization of meteorological objects, triboelectric textile fabrication, manufacturing of capacitive actuators and piezoelectric sensitive equipment. Metal structures, cellulose nano-crystals, and polarized tetra-fluoro-ethylene (TrFE) are some of the elements explored for such actuator and sensor-based applications (Ramanathan et al., 2022, Lee et al., 2021, Zhang et al., 2020). Some researchers have outlined the acceptable magnetic stimulus-based shape recovery, 4D properties in thermoplastic sheets, films, coils, 3D printable strands for optical sensors, dual-band energy harvesting antennas, and biomedical antennas preparation (Kumar et al., 2021, Ali et al.2020, Lopes et al., 2018). The investigations performed on tuneable dielectric properties of polymer composites, flexible and wearable patch antennas of recycled plastics have outlined the concept of using 3D/4D printed smart antennas for engineering as well as biomedical applications (Guo et al., 2022, Jain et al., 2021). Similarly, the results obtained by some researchers for self-assembly, self-actuation, and self-healing properties in 3D printed thermoplastic composite matrices have indicated that the 4D printing of smart sensors, actuators, and antennas may be performed effectively for batch production of customizable products for biomedical, industrial and structural engineering applications (Kumar et al., 2022a, Singh et al., 2022, Sharma et al., 2022). The 4D capabilities have been also reported in polyvinyl chloride (PVC), polypropylene (PP), and secondary recycled biocompatible polymers like polyvinyl alcohol (PVA), polylactic acid (PLA) that possess acceptable programming features for 4D printing of biosensors (Ranjan et al., 2021, Kumar et al., 2022b).

Based on reported literature on polymer matrix composites, it has been observed that PVDF thermoplastic has advanced applications as a composite for the repair and maintenance of non-structural cracks in heritage structures (Kumar et al., 2022c, Kumar et al., 2022d, Kumar et al., 2022e). Many research articles have outlined various applications of different PVDF composite matrices in different disciplines (like: engineering, medical, robotics, energy harvesting, self-healing, piezoelectricity, shape memory, electromagnetic induction shielding, etc.). Table 7.1 shows the list of various key terms/properties explored for PVDF thermoplastics and their composites along with their relevance scores (as per the Web of Science database for the past 20 years) for desired actuation-based sensor and antenna applications. Out of 889 research articles on polymer composite-based sensors, actuators, and antennas, 33 most relevant keywords were listed.

The literature shown in Table 7.1 is clustered in Figure 7.1 as a web of highly investigated key terms for PVDF and its composite for research areas like plastic

---

TABLE 7.1

List of Key Terms/Properties Investigated for PVDF and Its Composites

S. No.	Term	Studies Reported	Relevance Score
1	3D printed Structure	4	3.4939
2	Actuator	8	1.3025
3	Architecture applications	8	1.9197
4	Barium Titanate reinforcement	9	0.4084
5	Binder materials	4	1.4908
6	Carbon Nano-tubes	14	0.3658
7	Comparison of blending processes	6	0.6141
8	Composite Film	6	0.6223
9	Deposition	7	0.5209
10	Energy Harvesting	9	0.5472
11	Piezoelectric Polymer	6	0.6663
12	Piezoelectric PVDF	4	1.5607
13	Dielectric Property	5	1.0216
14	Differential Scanning Calorimetry	7	1.0936
15	Thermal properties	10	0.6946
16	Filament Fabrication	6	1.0685

(Continued)

TABLE 7.1 (*Continued*)
List of Key Terms/Properties Investigated for PVDF and Its Composites

S. No.	Term	Studies Reported	Relevance Score
17	Flexibility	10	0.4998
18	Frequency	6	0.6836
19	Graphene	9	0.3274
20	High Beta Phase Content	4	0.6313
21	High Sensitivity	5	0.9993
22	Material Extrusion	4	1.3866
23	Composite Matrix	13	0.6003
24	Mechanical Property	15	0.7437
25	Piezoelectric Coefficient	8	0.8842
26	Piezoelectric Device	8	1.0306
27	Polymer Programming	6	1.8957
28	Porosity	8	0.5114
29	Self Healing polymers	9	0.5011
30	Smart polymer composite	8	0.4239
31	Vibration sample magnetometry	18	0.7534
32	Voltage effect on conductivity	17	0.4931
33	Wide Range signals	4	0.6204

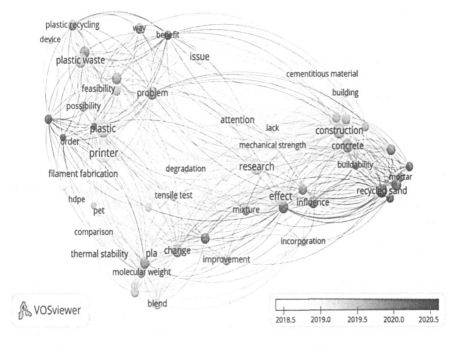

FIGURE 7.1 Web of key terms investigated for recycling and additive manufacturing (AM)-based applications of PVDF and its composites (as per Web of Science database).

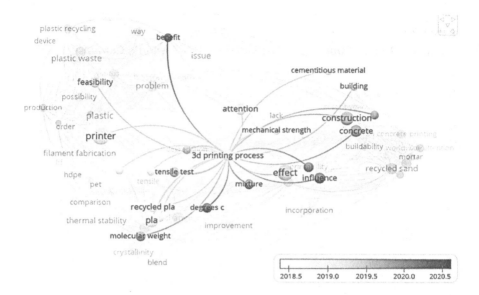

FIGURE 7.2   Research gap in a web of key terms for use of recycled plastic in 3D printing and structures.

waste recycling, 3D printing, construction, and manufacturing. Open-source software, VOS viewer has been used to obtain the cluster shown in Figure 7.1. Based on this, Figure 7.2 highlights the gap in the literature that electro-active, magnetic, and piezoelectricity-based 4D properties of PVDF composites are least reported for the 3D/4D printing processes that are linked with the development of smart sensors and antennas for monitoring the health of heritage buildings.

7.2   RESEARCH GAP

Some previous studies highlighted the preparation of PVDF composite reinforced with graphene and Mn-doped ZnO for self-actuation and PVDF-CaCO$_3$ composite with self-healing properties for the repair of cracks in heritage structures. The dual-material 3D printing of these composites also has been explored for controlling the weathering effect of heritage buildings. But hitherto little has been reported on the development of a 3D printable thermoplastic-based actuator (with 4D properties) for non-structural crack repairing, sensing, and online health monitoring of heritage structures by using the antenna-based applications as a smart actuator. This chapter highlights the fabrication of PVDF polymer composites-based dual-material 3D printed crack repair solution (for heritage structures) that also possesses the capabilities of a sensor to monitor the health of a crack (post-repair). The electrical and radio-frequency characterization of 3D/4D printed composite structure highlighted that the sensing properties of such functional prototypes may be used effectively to fabricate smart antennas for sensing damage to heritage structures and perform preventive maintenance by utilizing the online health monitoring features of such composites.

7.3   4D PRINTING OF SMART ACTUATOR
WITH DUAL MATERIAL 3D PRINTING

The Grade III heritage structure (as per the central public works department (CPWD) was studied for the development of a thermoplastic composite-based smart repairing solution by 3D printing. Figure 7.3 shows one of the damaged sites in the structure that was facing the removal of binding material due to weathering effect. Figure 7.4 shows the debris of the same site collected to prepare the composite for 3D printing by reinforcing the fine debris in a polymer matrix.

The cryogenic ball milling of debris was performed to obtain fine particles of $CaCO_3$ (raw form of debris). The $CaCO_3$ particles were blended in PVDF thermoplastic matrix to obtain 3D printable debris reinforced filament strand. The experimental studies performed on mechanical blending and chemical-assisted mechanical blending highlighted that chemical-assisted processing of PVDF-6%$CaCO_3$ composite resulted in better non-conducting (NC) type composite for 3D printing applications as the acceptable 4D properties based on chemical stimulus for shape recovery and programming features were obtained (Kumar et al., 2022d). To prepare conducting (C) type composite, chemical-assisted mechanically blended PVDF-6%graphene-3%Mn doped ZnO composition/proportion outlined good electrical, magnetic, and sensing properties for 3D printing applications (Kumar et al., 2022e). The dual-material 3D printing of alternative NC and C compositions/proportions resulted in the fabrication of 4D printed smart actuator strips with acceptable self-elongation and self-contraction of ~198 ppm/cm^2 under the applied electric field of 10 kV. The 3D printed customizable strip possesses an acceptable corrosion rate of ~2.8 mm/yr for its application in repairing non-structural cracks (Kumar et al., 2022b). It should be noted that the C-NC-C 3D printing fashion of the same strip-like prototype may

FIGURE 7.3   Wall of heritage structure facing degradation of binding material.

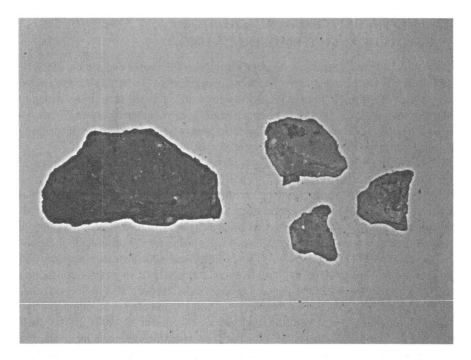

FIGURE 7.4   Debris collected from heritage site for composite preparation.

be used to induce sensor and antenna properties in the crack repairing solution for monitoring the health of repaired damage (crack).

7.4   SENSING-BASED ANTENNA APPLICATIONS OF USER-DRIVEN SMART SOLUTION

Based on previously reported sensing capabilities of thermoplastic composites for antenna applications by Singh et al., 2022, C-NC-C composite layers were 3D printed to fabricate a rectangular dual-material printed antenna substrate. The ring resonator test was applied on the composite substrate to obtain the resonating frequency (RF) of the 3D printed sensor. The rectangular substrate (having dimension 40×50×0.9 mm (length × breadth × thickness)) outlined the RF of 2.6 GHz from RF of 2.71 GHz when the piezoelectric effect was induced in the sample by bending it along the diagonals. This test indicated that miniaturization of a proposed 3D printed sensor may be performed to obtain the optimized design and dimension for its antenna application in the industrial, scientific and medical (ISM) band.

7.4.1   DESIGNING OF PATCH ANTENNA

The patch of the antenna was designed based on the specifications such as the RF, patch length, width, and effective dielectric constant(ϵ_{eff}) for composite material. For health monitoring of the repaired non-structural crack in heritage structure (within the ISM band), the patch antenna was designed for RF of 2.45 GHz with 0.9 mm substrate thickness. Equations (7.1–7.4) reported in the literature were used to calculate the dimensions of the patch as a micro patch antenna (MPA) (Singh et al., 2022).

Equation (7.1) given below gives the width of the patch of antenna:-

$$W = \frac{c}{2f_r}\sqrt{\left(\frac{2}{\epsilon_r+1}\right)} \tag{7.1}$$

Here,
W=width of the patch,
f_r=resonant frequency,
ϵ_r=dielectric constant
The effective dielectric constant (ϵ_{eff}) is given by Equation 7.2.

$$\epsilon_{eff} = \left[\frac{\epsilon_r+1}{2}\right] + \left[\left(\frac{\epsilon_r-1}{2}\right)\left(1+12\frac{h}{W}\right)^{-0.5}\right] \tag{7.2}$$

Here,
h = thickness of the substrate
The value of ΔL, the fringe factor, needs to be calculated to find the length of the patch.

$$\Delta L = 0.42h\left[\frac{\left(\epsilon_{eff}+0.3\right)\left(\frac{w}{h}+0.264\right)}{\left(\epsilon_{eff}-0.258\right)\left(\frac{w}{h}+0.8\right)}\right] \tag{7.3}$$

The length of a patch is calculated as:-

$$L = \frac{C}{2f_r\sqrt{\left(\epsilon_{eff}\right)}} - 2\Delta L \tag{7.4}$$

The dimensions of the patch are obtained in Equations (7.5) and (7.6).

$$W_s = W + 6h \tag{7.5}$$

$$L_s = L + 6h \tag{7.6}$$

Here, W_s and L_s are width and length of the substrate respectively and h is the thickness of the substrate. Table 7.2 shows the calculated dimensions for length and width of patch for 4 different sizes of 3D printable antenna substrates for the proposed condition monitoring applications (using Equations 7.1–7.6).

7.4.2 SIMULATION OF ANTENNA

Further simulation for the patch antenna was performed by using the high-frequency structure simulator (HFSS) 15.0.3 software package to simulate the patch antennas

TABLE 7.2
Dimensions of Patch for Antennas at RF 2.45 GHz

Substrate	Width of a Patch (mm)	Length of a Patch (mm)
1	37	29
2	38	29.5
3	39	30.5
4	39.16	30.9

for resonating frequency of 2.45 GHz. The designed sensors were then simulated on HFSS software for RF and the gain of the sensors. For sample 1 (as per Table 7.2) it was observed that it resonates at 2.4 GHz with a reflection loss of −7.22 dB (as shown in Figure 7.5) indicating that it did not fulfill the condition for the patch antenna (i.e. reflection loss should be < −10 dB).

Further, the gain for sample 1 as a sensor was simulated (Figure 7.6) which comes out to 3.44 dB at 2.40 GHz.

The simulated results for sample 2 show RF of 2.40 GHz at a reflection loss of −11.1543 dB which shows its acceptable antenna-based sensing characteristics (with an increase in patch size, width 38 mm, length 29.5 mm) as shown in Figure 7.7. The simulated gain was observed at 3.33 dB (Figure 7.8). The results here indicated a decrease in gain as compared to sample 1.

Further simulation-based results for sample 3 indicated the RF of 2.40 GHz at a reflection loss of −10.96 dB which shows its sensing characteristics with a slight increase in the size of a patch as shown in Figure 7.9. The simulated gain of 3.37 dB is shown in Figure 7.10.

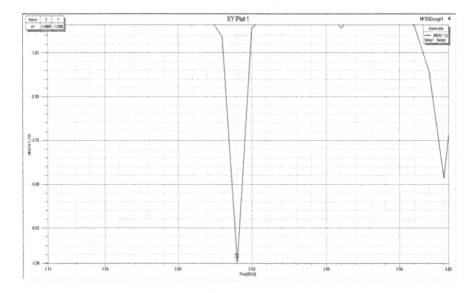

FIGURE 7.5   Graph for RF versus S_{11} parameters for sample 1.

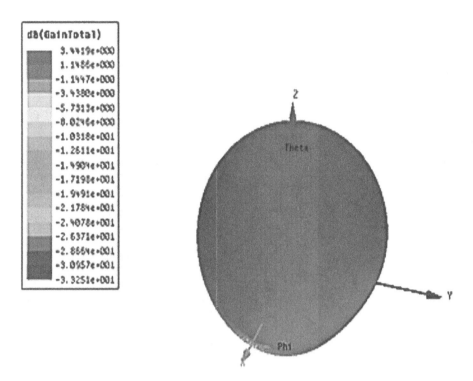

FIGURE 7.6 Simulated gain for sample 1 (Dimensions: 37×29 mm (width × length)).

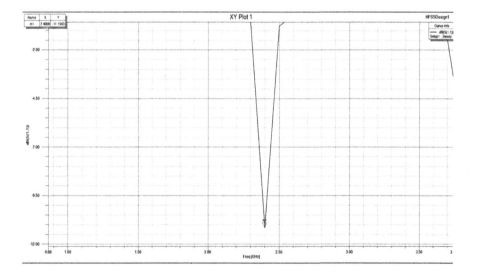

FIGURE 7.7 Graph for RF versus S_{11} parameters for sample 2.

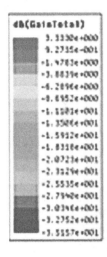

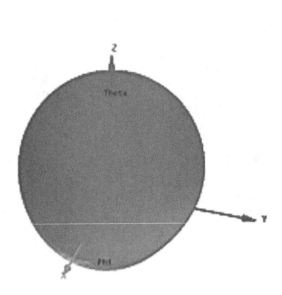

FIGURE 7.8 Simulated gain (sample 2).

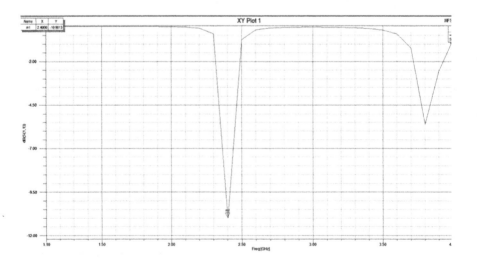

FIGURE 7.9 Graph for RF versus S_{11} parameters for sample 3.

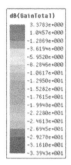

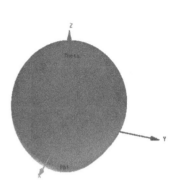

FIGURE 7.10 Simulated gain (sample 3).

Similarly, the results obtained for simulations of sample 4 for 2.40 GHz RF at a reflection loss of −13.16 dB show its sensing characteristics with an increase in patch size (width=39.16 mm, length=30.9 mm) (Figure 7.11). The simulated gain of 3.28 dB is shown in Figure 7.12.

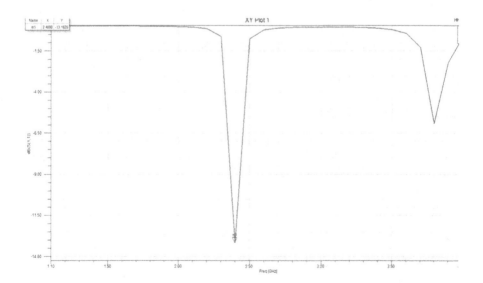

FIGURE 7.11 Graph for RF versus S_{11} parameters for sample 4.

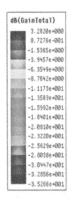

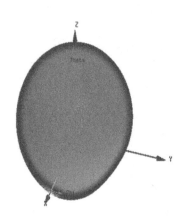

FIGURE 7.12   Simulated gain (sample 4).

7.5   SUMMARY

The investigations performed on dual-material 3D printed 4D capable smart-customizable repair solutions of heritage structures for sensor or antenna-based online health monitoring of repaired cracks outlined that the acceptable self-elongation and self-contraction like actuator features of such thermoplastic composites may be used as a novel approach to fabricate smart sensors and antennas for monitoring the weathering and crack propagation of repaired non-structural cracks of heritage structures. The proposed 4D capable smart dual composite material-based sensor may be used within ISM band of antenna applications for online condition monitoring.

ACKNOWLEDGMENT

The authors are thankful to the Department of Science and Technology (DST) research project DST-SHRI for providing the financial support for this work (File No. DST/TDT/SHRI-35/2018)

REFERENCES

Ali, T. G., Bai, X., Xu, L. J. Dual-band energy harvesting antenna based on PVDF piezo-electric material. In 2020, 9th Asia-Pacific Conference on Antennas and Propagation, (APCAP) 2020, Aug 4 (pp. 1–2), IEEE

Guo, Y., Liu, S., Wu, S., Jie, X., Pawlikowska E., Bulejak W., Szafran, M., Rydosz, A., Gao, F. Enhanced tunable dielectric properties of Ba0.6Sr0.4TiO3/PVDF composites through dual-gradient structural engineering, Journal of Alloys and Compounds, 2022, 166034, https://doi.org/10.1016/j.jallcom.2022.166034

Jain, C., Dhaliwal, B. S., Singh, R. Flexible and Wearable Patch Antennas Using Additive Manufacturing: A Framework. Ref Mod. Mater. Sci. Mater. Eng. 2021, https://doi.org/10.1016/B978-0-12-820352-1.00093-6

Kumar, V., Singh, R., Ahuja, I. S. 3D-printed innovative customized solution for regulating weathering effect on heritage structures. Materials Letters, 2022c Jun 22:132717

Kumar, V., Singh, R., Ahuja, I. S. On 3D printing of electro-active PVDF-Graphene and Mn-doped ZnO nanoparticle-based composite as a self-healing repair solution for heritage structures, Proceedings of the Institution of Mechanical Engineers, Part B: Journal of Engineering Manufacture. 2022e; 236(8):1141–1154

Kumar, V., Singh, R., Ahuja, I. S. On Rheological, Thermal, Mechanical, Morphological, and Piezoelectric Properties and One-Way Programming Features of Polyvinylidene Fluoride–CaCO$_3$ Composites, Journal of Materials Engineering and Performance, 2022d:1–5

Kumar, S., Singh, R., Singh, T. P., Batish, A. On 3D-printed multi-blended and hybrid-blended poly (lactic) acid composite matrix for self-assembly. In 4D Printing, 2022a (pp. 1–15). Elsevier

Kumar, R., Singh, R., Kumar, V., Kumar, P., Prakash, C., Singh, S. Characterization of in-house-developed Mn-ZnO-reinforced polyethylene: a sustainable approach for developing fused filament fabrication-based filament. Journal of Materials Engineering and Performance, 2021; 30(7):5368–5382

Kumar, V., Singh, R., Ahuja, I. S. Secondary recycled polyvinylidene–limestone composite in 4D printing applications for heritage structures: Rheological, thermal, mechanical, spectroscopic, and morphological analysis. *Proceedings of the Institution of Mechanical Engineers, Part E: Journal of Process Mechanical Engineering*. 2022b. doi:10.1177/09544089221104771

Lee, J. E., Shin, Y. E., Lee, G. H., Kim, J., Ko, H., Chae, H. G. Polyvinylidene fluoride (PVDF)/cellulose nanocrystal (CNC) nanocomposite fiber and triboelectric textile sensors. Composites Part B: Engineering. 2021; 223:109098

Lopes, A. C., Gutiérrez, J., Barandiarán, J. M. Direct fabrication of a 3D-shape film of polyvinylidene fluoride (PVDF) in the piezoelectric β-phase for sensor and actuator applications. European Polymer Journal. 2018; 99:111–116

Ramanathan, A. K., Gingerich, M. B., Headings, L. M., Dapino, M. J. Metal structures embedded with piezoelectric PVDF sensors using ultrasonic additive manufacturing. Manufacturing Letters. 2022; 31:96–100

Ranjan, N., Kumar, R., Singh, R., Kumar, V. On PVC-PP composite matrix for 4D applications: Flowability, mechanical, thermal, and morphological characterizations. Journal of thermoplastic composite materials. 2021

Sharma, R., Singh, R., Batish, A. PVDF-graphene-BaTiO3 composite for 4D applications. In 4D Printing 2022 (pp. 103–119). Elsevier

Singh, R., Kumar, S., Kumar, R. On dual/multi-material composite matrix for smart structures: a case study of ABS-PLA, HIPS-PLA-ABS. In 4D Printing 2022 (pp. 89–101). Elsevier

Singh, R., Kumar, S., Singh, A. P., Wei, Y. On comparison of recycled LDPE and LDPE–bakelite composite-based 3D-printed patch antenna. Proceedings of the Institution of Mechanical Engineers, Part L: Journal of Materials: Design and Applications. 2022; 236(4):842–856. https://doi.org/10.1177/14644207211060465

Zhang, Q., Liu, S., Luo, H., Guo, Z., Hu, X., Xiang, Y. Hybrid capacitive/piezoelectric visualized meteorological sensor based on in-situ polarized PVDF-TRFE films on TFT arrays. Sensors and Actuators A: Physical. 2020 Nov 1; 315:112286

8 Case Study on the Development of Polymer Composite for Sensors, Actuators, and Antennas

Rupinder Singh and Abhishek Barwar
National Institute of Technical Teacher
Training and Research, Chandigarh

CONTENTS

8.1 INTRODUCTION

Optimization of electronic devices such as sensors, actuators, and antennas was always an important concern for design engineers (Saleem I., 2012). Gain enhancement, increase in efficiency, bandwidth, miniaturization, specific absorption rate (SAR) adjustments, etc. were some of the important parameters which need to be

DOI: 10.1201/9781003194224-8

dealt with primarily (Ahmed et al., 2012). Microstrip antennas were devices that consist of a patch, ground, and substrate grown up in the early 1970s, these have numerous applications in a no. of areas working in a microwave system (Menzel and Grabherr, 1991). Changing the substrate material and its thickness can cause a change in system performance as the dielectric constant is a material property and comes under one of the effective parameters while designing an antenna (Paul et al., 2015). Generally, the materials whose dielectric constant (ε_r) lies within the range ($2 \leq \varepsilon_r \leq 12$) are found suitable for the substrate material (Zaidi and Tripathi, 2014). A wide variety of dielectric materials with favorable mechanical, electrical, and thermal properties exist to be used for designing an antenna for both planar and conformal configurations such as RT Duroid, FR4, PDMS, Bakelite, LDPE, ABS, etc (Keith and James, 1981). PLA is a naturally derived polymer with attractive features such as biocompatible, biodegradable, and built-in mechanical, and electrical properties proven to be a good thermoplastic polymer for the preparation of a composite (Rasal and Hirt, 2009, Ray and Okamoto, 2003). CS was found to be a biomaterial that was used as a reinforcing material, it possesses great anti-bacterial activity and biocompatibility and was used along with other polymers to prepare a composite, apart from that to attain bone-like properties. HAP was an excellent biomaterial that along with CS enhance the tissue regeneration rate (Kawakami et al., 1992). Substrate made up of PLA shows more availability of radiation power as compared to other reference materials because of the minimum return losses and minimum reflection of power (Ullah and Flint, 2014). The use of smart implantable devices such as pacemakers, and swallowable pills enable the health monitoring and diagnosis of a patient remotely by placing the signal radiating device inside the body of a patient which is connected with a receiver unit (Kaur et al., 2015). Wireless sensors for biomedical applications attract a lot of researchers nowadays due to their capability of measuring various physiological parameters remotely (Sukhija and Sarin, 2017). Medical Implant Communication System (MICS) and industrial scientific and medical (ISM) bands were approved for biomedical applications of electronic devices operatable in the frequency bands 402–405 MHz and 2–2.5 GHz respectively (Smith EK., 1999). A literature review reveals that a lot of polymer materials exist for sensors, actuators, and antennas, but hitherto very less has been reported on the use of PLA-based composite specifically for implantable devices.

To identify the research gap, a detailed literature review based on the bibliographic study has been carried out on the web of knowledge database platform. Since this case study focuses on the preparation of polymer composite for sensors, actuators, and antennas therefore the terms- "actuator, polymer matrix, bio-sensor, and preparation" were searched collectively on the web of science database, and as a result, the research record of past two decades has been obtained that can be exported in the form of a text file. The literature data was further processed in the Visualization of science (VOS) viewer networking diagram tool, where it was found that out of 2507 terms, 67 meet the threshold value when the minimum occurrence of a term is restricted to 5. For each of these 67 terms, a relevance score has been evaluated, and based on that only 60% most relevant were retained. And finally, the network diagram was developed based on the mutual research that happened between the above four terms. The obtained diagram contains three clusters represented by different color codes i.e., red, blue, and green, these carry out the information about the independent research reported in different areas. Moving ahead, Figure 8.1 (a–d) tells about the work that happened for the particular node by highlighting keywords-actuator, the polymer matrix, bio-sensor, and preparation of composites respectively. Figure 8.1 tells about different works reported regarding the fabrication of polymer matrix, actuator, and sensors, utilizing PLA as a material but hitherto little has been identified about the use of PLA-based composite for developing sensor, actuator, and antenna operating in ISM band.

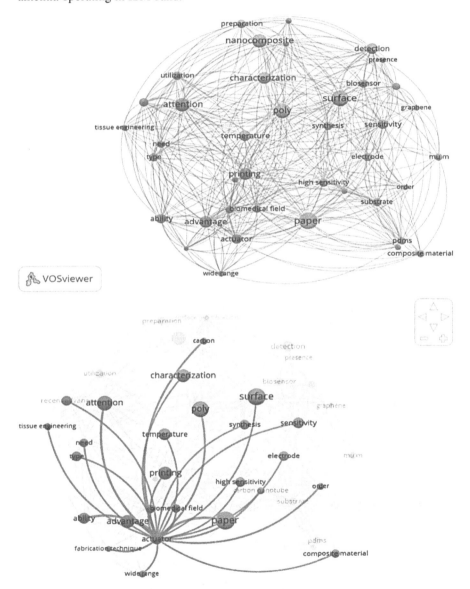

FIGURE 8.1 Networking diagram (a) to analyze the research gap based on terms: actuator (b), polymer matrix. (*Continued*)

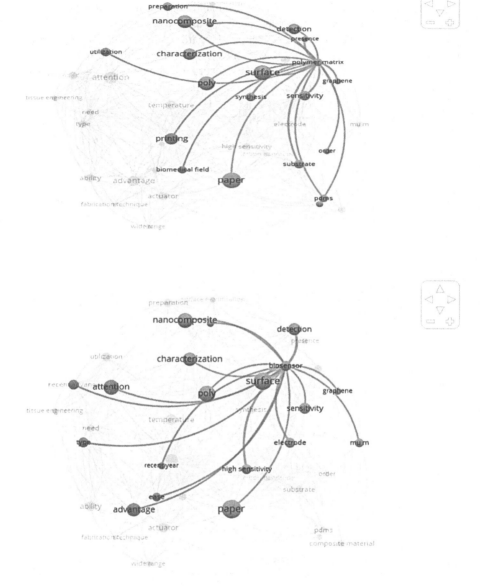

FIGURE 8.1 (*Continued*) Networking diagram (c), bio-sensor (d), and preparation.

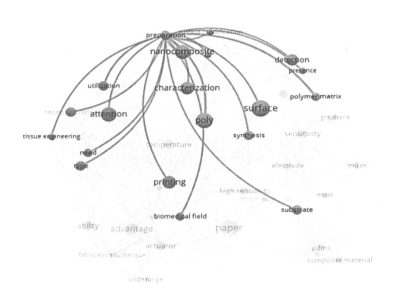

FIGURE 8.1 (*Continued*)   Networking diagram (e) with nodal point preparation.

Table 8.1 contains the information on the no. of occurrence of a term in the literature of the past two decades and a corresponding relevance score for each term was calculated. The total no. of terms (60) that occurred in both Figures 8.1 and 8.2 were combinedly represented in Table 8.1, and these networking diagrams were developed by using the terms mentioned in Table 8.1.

TABLE 8.1

Relevance Score of the Terms That Contribute to the Formation of Linkage Diagram

S.No.	Term	Occurrences	Relevance Score
1	Ability	8	0.9982
2	Actuator	10	0.6388
3	Advantage	10	0.627
4	Attention	12	0.6506
5	Biomedical field	6	0.5319
6	Biosensor	7	1.086
7	Carbon	6	0.6228
8	Carbon nanotube	6	0.5801
9	Characterization	10	0.3428
10	Composite material	6	2.3623
11	Detection	9	1.2227
12	Ease	6	0.5011
13	Electrode	7	0.5141
14	Fabrication technique	5	1.1886
15	Graphene	5	1.7774
16	High sensitivity	8	0.4162

(Continued)

TABLE 8.1 (*Continued*)
Relevance Score of the Terms That Contribute to the Formation of Linkage Diagram

S.No.	Term	Occurrences	Relevance Score
17	MUM	6	2.0908
18	Nanocomposite	11	0.9483
19	Need	6	0.5659
20	Order	5	0.7816
21	Paper	14	0.3448
22	PDMS	6	2.6452
23	Poly	12	0.2787
24	Polydimethylsiloxane	5	2.9756
25	Polymer matrix	6	1.4927
26	Preparation	7	1.5151
27	Presence	5	1.526
28	Printing	10	0.5758
29	Recent advance	8	0.9119
30	Recent year	5	0.4103
31	Sensitivity	9	0.5543
32	Substrate	7	0.841
33	Surface	13	0.5404
34	Surface modification	5	1.1394
35	Synthesis	8	0.4238
36	Temperature	9	0.4872
37	Tissue engineering	6	1.5399
38	Type	7	0.8036
39	Utilization	8	1.3285
40	Wide range	5	1.2188
41	Bandwidth	5	0.4172
42	Circuit board	3	1.3117
43	Dbi	3	1.3539
44	Development	4	0.256
45	Efficiency	4	1.1591
46	Electromagnetic performance	3	0.6905
47	Epsilon	3	1.5626
48	Fabrication	7	1.6172
49	Function	3	1.0492
50	Loss tangent	4	0.9752
51	Low cost	3	1.2789
52	Microstrip patch	4	0.3229
53	Microstrip patch antenna	6	1.1021
54	Morphology	3	2.1701
55	Polymer	7	0.6545
56	Prototype	7	0.9227
57	Radiation efficiency	3	0.9067
58	Radiation pattern	8	0.6573
59	Use	3	1.3412
60	X band	3	0.2511

Figure 8.2 (a) represents the bibliographic analysis based on the keyword-microstrip patch antenna, it contains 4 clusters represented by color codes-red, blue, yellow, and green. The figure tells about the work associated with microstrip patch antenna in the past two decades i.e. when highlighting any particular node, its linkage with other nodes appeared if mutual research has been reported between the respective nodes. Figure 8.2 (b) represents the gap analysis by highlighting the term microstrip antenna, and it was not connected with the low-cost node which represents the lag in research between the nodes. Therefore, a low-cost polymer composite has been developed in this study to overcome the gap.

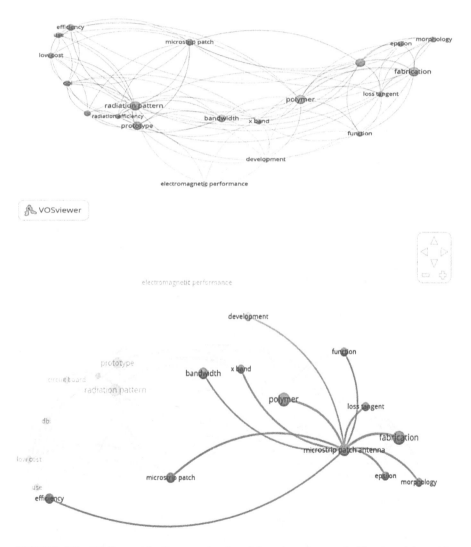

FIGURE 8.2 Bibliographic diagram developed based on keyword: (a) microstrip patch antenna, (b) gap analysis.

8.2 METHODOLOGY

In this case study, the preparation of polymer composite has been ascertained (Figure 8.3) to develop a sensor, actuator, and antenna by utilizing PLA with some proportions of HAP and CS. Initially, to identify the flow behavior of the material, the MFI has been carried out of the material (PLA-HAP-CS) with different compositions i.e., 89-8-3; 90-8-2; and 91-8-1. Further, based upon MFI values, the

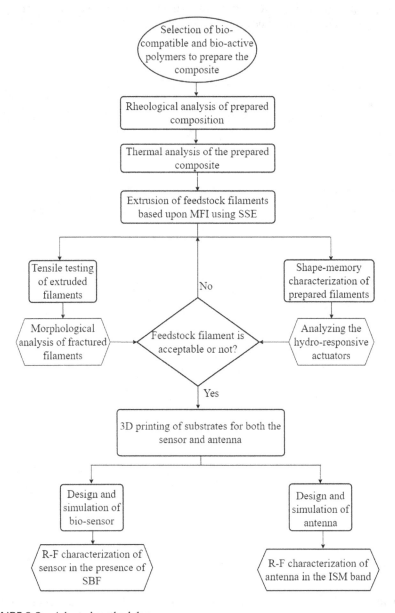

FIGURE 8.3 Adopted methodology.

composition (90-8-2) was selected for further investigations. For the extrusion of the selected composition, a set of filaments has been prepared on single screw extruder (SSE). To check the mechanical behavior of the material under loading conditions, tensile testing of the material has been carried out on universal testing machine (UTM). By performing the fracture analysis, the morphological features were extracted from the fractured site by using the scanning electron microscopy (SEM) image and energy dispersive spectroscopy (EDS) data. Going on, to understand the fracture behavior, Fourier transform infrared spectroscopy (FTIR) has been carried out to identify the bond characteristics at the fractured site. For analyzing the thermal behavior of the composite, a thermal analysis has been ascertained on a differential scanning calorimetry (DSC) setup. In this study, the composite was prepared for biomedical application, and hence the shape memory behavior of the material has been analyzed by giving the stimulus in the form of moisture (hydro), and based upon that the response of the material has been recorded. After performing different analyses, it was decided whether the prepared composite was desirable or not, if yes then by using the feedstock filament, the substrates of the bio-sensor, as well as antenna, have been prepared on FDM. Based on the theoretical values of dielectric constant and loss tangent of the substrate material, the sensor and antenna have been designed and simulated in the HFSS software and by performing a ring resonator test, the dielectric properties of the material have been ascertained. Finally based on the simulated results, the antenna and sensor have been developed and the RF characteristics of the bio-sensor have been ascertained by giving the exposure of simulated body fluid to the surface of the sensor, and similarly, the RF behavior of the antenna has been analyzed by selecting the operating frequency in the ISM band.

8.3   SELECTION AND PREPARATION OF MATERIAL

To prepare a polymer composite for the development of sensors, actuators, and antennas specifically for biomedical applications, the materials should be bio-active. A lot of materials were explored by researchers for the fabrication of implantable biosensors as well as antennas. PLA was found to be a naturally derived polymer having outstanding biodegradability, and in addition to that, PLA also possesses the ability to be used in a variety of structures, and geometry for a no. of biomedical applications (Ariyapitipun et al., 1999). HAP was one of the bio-material generally utilized for bone recovery or bone renovation because of its organic action and biodegradable nature (Burg et al., 2000). It was generally found in the natural bone along with the collagen fibers and hence found to be a suitable material for the preparation of a composite for the fabrication of a sensor or antenna (Kikuchi et al., 2001). In this study, HAP with 150–190 mesh size was utilized as a filler material for the preparation of polymer composite. Other than that, for attaining anti-bacterial, excellent holding properties with the surrounding tissues, the use of CS as a biomaterial has been reported (Madihally and Mathew, 1999). CS along with HAP formed such a composite that offers the best biocompatibility, enhanced osteoconductivity, and improved mechanical properties (Wan et al., 1998). CS with deacetylation greater than 90%, and mesh size 35–40 was used in this work. The bio-materials were first heated inside a hot oven for 2 hours to make these moisture free and then mixed properly using a ball mill setup operating at 40 rpm for 4 hours (Ranjan et al., 2019).

8.4 EXPERIMENTATION

8.4.1 RHEOLOGICAL INVESTIGATION

The rheology of the prepared composition has been ascertained by performing the MFI as per ASTM D1238. This standard ensures the printability of the thermoplastic polymers as well as their composites generally. Studies suggest the use of CS-based nanofibers suitable for the fabrication of biosensors (Jayakumar et al., 2010). To analyze the effect of varying the proportions of HAP and CS on the flow behavior of PLA-based composite, three sets of compositions were prepared (i.e., 89-8-3; 90-8-2; and 91-8-1). The compositions were then tested on an MFI tester by setting the temperature at 185°C and applying an external load weighing 2.160 kg.

8.4.2 DSC ANALYSIS

The thermal behavior of the PLA-based composite has been analyzed by using the DSC setup (Make: Mettler Toledo, Switzerland) having a maximum temperature limit of 500°C to identify the endothermic and exothermic reactions that occurred during the experiment (Figure 8.4). During the sample preparation stage, the composite sample was weighed (i.e., 2.3 mg) initially and placed into a sample pan. The sample along with the reference pan put onto a thermo-electric disk which was

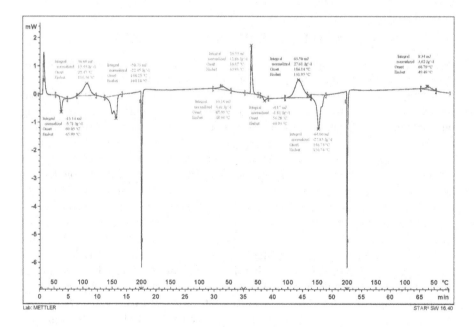

FIGURE 8.4 Thermal stability analysis of the composite material.

enclosed inside a furnace, therefore transfer of heat will take place from the disk to the pan at a linear rate. The physical or chemical changes measured within the sample during the heating process indicate the transfer of energy to or from the sample. While performing the thermal stability test of the composite, the segment gas was chosen as air with a flow rate of 50 ml/min, whereas the heating rate was fixed to 10°C/min (Singh et al., 2022).

8.4.3 EXTRUSION OF FEEDSTOCK FILAMENTS

For the fabrication of the sensor, actuator, and antenna via the 3D printing method (FDM), the feedstock filament has been required as input material. Based upon MFI results, the composition (90-8-2) was found suitable for 3D printing of sensor or antenna. During the extrusion process, the material was fed inside the hopper and extruded out through a die of 1.75 mm. Once the material comes out through SSE in the wire form, a filament wrapping arrangement has been used to continuously monitor the diameter of the wire to ensure the uniformity required for processing through a 3D printer. The filaments were extruded using SSE (Felfil, Italy) at T=180°C and N=6 rpm.

8.4.4 MECHANICAL BEHAVIOR

To identify the mechanical properties of the prepared composite, the tensile testing of the wire samples has been ascertained on UTM (Shanta Engineering, Pune, India) having a maximum load capacity of 5000 N. The UTM machine was equipped with its software to set input parameters such as wire diameter (1.75 mm), strain rate (30 mm/min), gauge length, etc. The grip separation length was considered as 50 mm while testing the sample. Once the sample got fixed in between the fixed and movable jaws of the UTM, the sample gets strained at a constant rate and the response got recorded in the form of a load vs deformation curve.

8.4.5 MORPHOLOGICAL INVESTIGATION

To understand the morphology of the mechanically tested wire samples, the SEM photomicrographs of the fractured site have been captured at different magnifications. For analyzing the presence of different elements in the composite, EDS (based on SEM image) has been performed. Since the composite has been prepared for sensor or antenna fabrication, therefore the surface texture plays a significant role in the output signal. To understand the surface features, average roughness parameters of the cross-section were calculated using Gwyddion software. The bonding behavior of the composite has been analyzed by performing the FTIR of the filament samples.

8.4.6 HYDRO-RESPONSIVE ACTUATOR

Since the sensor or antenna developed in this study has to be meant for biomedical applications, therefore adding a hydro-responsive property to the final product makes it a smart device whose behavior changes with the exposure to a hydro-based solution. These kinds of actuators play a significant role in biomedical science as they have stimulus-responsive capability (Banerjee et al., 2018). In this study, to check the hydro-response of the material, the samples having a length of 5-10 mm were dipped into the water container at 40°C for 24 hours and then removed to observe the volumetric changes that occurred in the samples.

8.4.7 DESIGN AND SIMULATION OF THE SENSOR

After analyzing the different properties of the polymer-based composite, the designing stage of the sensor appeared in which the bio-sensor has been designed in the HFSS software. It comprises three layers (i.e., ground plane, radiating surface, and the intermediate layer named substrate) and works on the principle of a parallel plate capacitor. Since it has to be designed for an operating frequency of 2.45 GHz, the thickness of the substrate was considered as 1 mm whereas the theoretical value of ε_r was taken as 2.6, and the dimensions of the bio-sensor have been analyzed by following the calculations (Equations 8.1–8.6) (Singh et al., 2022);

$$2\pi R = n\lambda_g \tag{8.1}$$

$$\lambda_g = \frac{c}{\sqrt{\varepsilon_{eff}}} \times \frac{1}{f_r} \tag{8.2}$$

$$\varepsilon_{eff} = \left[\frac{\varepsilon_r + 1}{2} \right] + \left[\left(\frac{\varepsilon_r - 1}{2} \right) \left(1 + 12 \frac{h}{w} \right)^{-0.5} \right] \tag{8.3}$$

$$L = \frac{\lambda_g}{4} \tag{8.4}$$

$$R_i = R - \frac{w}{2} \tag{8.5}$$

$$R_o = R + \frac{w}{2} \tag{8.6}$$

Where R is the mean radius of the ring; λ_g is the guided wavelength; ε_{eff} is the effective dielectric constant of the material; f_r is the resonant frequency; ε_r is the dielectric constant of the material; h is the height of the substrate; w is the width of the ring; R_i is the inner radius of the ring, and R_o is the outer radius of the ring.

After designing the sensor, the simulation was performed using HFSS software and the RF behavior of the sensor was recorded in the form of S_{21} vs f_r. Going ahead, the fabrication of bio-sensor has been ascertained on FDM setup with printing parameters: layer thickness-1 mm, raster angle-45°, and infill density-100%. Since the sensor was developed for biomedical applications therefore the RF characteristics were obtained by giving the exposure to simulated body fluid (SBF) (Figure 8.5).

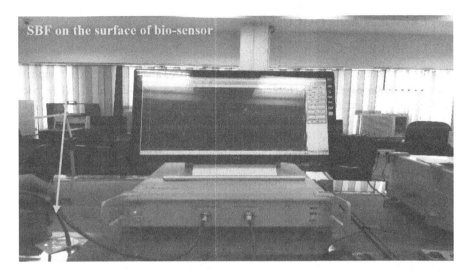

FIGURE 8.5 Bio-sensor tested onto VNA set up under the influence of SBF.

8.4.8 DESIGN AND FABRICATION OF MICROSTRIP PATCH ANTENNA

The antenna was designed (Figure 8.6) based upon the dielectric properties of the composite (i.e., ε_r and tan δ) which was calculated from the output (i.e., S_{21} and f_r) obtained through the ring resonator test. Apart from that, the antenna has been designed for the resonating frequency of 2.45 GHz lying in the ISM band, and the calculated dimensions of the rectangular patch were L=33.35 mm and W=41.66 mm. From the fabrication point of view, similar printing parameters were selected that were used for the fabrication of the sensor, and hence the substrate has been printed on FDM whereas the conducting patch and ground plane were designed using adhesive copper tape having a thickness of 0.08 mm. The performance of the antenna has been analyzed by comparing the simulated results with the real-time results i.e., the response was recorded in the form of return loss (S_{11}) vs f_r.

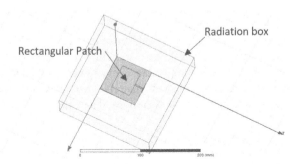

FIGURE 8.6 3D model of patch antenna designed using HFSS software.

8.5 RESULTS AND DISCUSSION

8.5.1 MELT FLOW CHARACTERISTICS

Based on the rheological analysis, the measured MFI values for selected compositions were 7.563, 11.327, and 11.997 g/10min for 89-8-3, 90-8-2, and 91-8-1 respectively. It was noticed that with the decrease in weight proportions of CS and increase of PLA weight proportion, the MFI value of the PLA-based composite has increased continuously. The higher value of MFI represents the enhanced rate of filament extrusion. Previous literature recommends the 90-8-2 composition to be used for the development of biomedical scaffolds and implants (Ranjan et al., 2019). Since the sensors and antenna were designed for an implantable device to be utilized for health monitoring of patients, therefore the 90-8-2 composition was used for the filament extrusion in this study.

8.5.2 THERMAL BEHAVIOR

The thermal stability of the prepared composite was analyzed by measuring the physical or chemical changes that occurred in the material during the energy transferred for two consecutive heating-cooling cycles. The temperature range for the composite was considered within the range of 25 to 200°C, and the heat was supplied and rejected at a rate of 10°C/min. Previous studies suggest that the first heating-cooling cycle is generally not considered for evaluation because of the presence of the foreign substance (Ranjan et al., 2019). During the 2nd heating cycle, two peaks were observed one endothermic (onset temp:106.14°C, endset:131.95°C) and the other was exothermic (onset temp:146.74°C, endset: 156.74°C). The first peak indicates the occurrence of glass transition, whereas the 2nd peak represents the cold crystallization of the material. During the 2nd cooling cycle, one endothermic peak (onset temp: 66.79°C, endset: 49.49°C) was observed representing the melting behavior of the material. The glass transition temperature of the composite was observed at 67°C.

8.5.3 TENSILE PROPERTIES

To ensure the mechanical strength of the polymer-based composite, the filament samples were strained on the UTM and different mechanical properties of the material have been obtained. The measured values of the mechanical properties were, ultimate tensile strength: 44.76 MPa, break strength: 37.49 MPa, peak strain: 0.023, modulus of toughness (MOT): 0.61 MPa, and Young's modulus: 2087.42 MPa. The above data indicates that the composite material possesses a higher value of peak strength which was suitable for the fabrication of implantable devices where high mechanical strength is desired. Apart from that, the composite material also showcases great stretchability which was utilized for the fabrication of conformal antennas.

8.5.4 SURFACE AND BONDING CHARACTERISTICS

The SEM images of the fractured samples were captured at 300X by maintaining a working distance of 9 mm and a scanning electron detector (SED) was used to

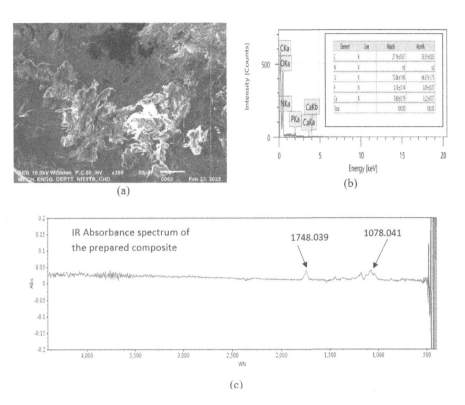

FIGURE 8.7 Surface characteristics of the fractured sample: (a) SEM image at 300X, (b) EDS data based on SEM image, (c) Bond characteristics of a fractured site.

generate the topography of the sample at 10 kV. Figure 8.7 highlights the uniform mixing (the fibrous structure in the image were the CS flakes) of the HAP and CS with PLA throughout the volume of the sample. The elemental composition was represented in the EDS graph, the presence of various elements in the composite with their mass % and atomic % was found to be very similar to that of actual materials. The bonding characteristics were obtained by performing the FTIR within wave no. range 500–4000 by analyzing the absorbance bands present in the spectrum. The first absorbance peak was observed at 1078.04 which represents the presence of the C-F stretch bond-forming aliphatic fluoro compounds. Moving ahead, another peak was obtained at 1748.03 which represents the presence of a strong C=O stretch bond belonging to carbonyl compounds which may be due to the bonding of methyl with the benzene ring (Nandiyanto et al., 2019). The enhanced peak strength of the composite may be due to the presence of a strong double bond.

8.5.5 ACTUATION BEHAVIOR

The actuation behavior has been recorded by providing the hydro-based stimulus to the filament sample. After providing the hydro-exposure to the sample for 24 hours, there will be an increase in the dimensions of the sample i.e., the length increases

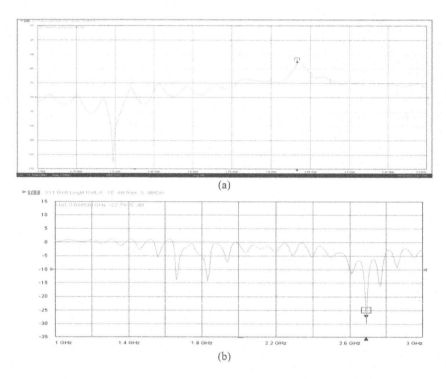

FIGURE 8.8 RF characteristics of the biosensor and patch antenna.

from 20.25 to 20.38 mm, whereas the diameter increases from 1.40 to 1.46 mm. The volumetric expansion of the samples represents the capability of the material to expand or contract with the effect of stimulus and hence can be employed for the development of smart devices for biomedical applications (i.e., implantable sensors or antenna).

8.5.6 RF Characterization of Bio-Sensor and Antenna

The RF characteristics of both the sensor and antenna have been obtained on the VNA setup by performing the initial settings i.e., the calibration of the ports (open, short, and through), selection of frequency sweep, and since both the devices were meant for ISM band, therefore, the frequency range for observing the RF behavior was selected as 1 to 3 GHz. As shown in Figure 8.8(a), the sensor resonates at 2.99 GHz with a reflection coefficient of −40.82 dB. The effect of SBF on the RF behavior was observed in a way that the sensor was resonating at a higher frequency range, therefore to make it resonate in the ISM band, the designing parameters need to be changed. Similarly, the RF characteristics for the antenna have been obtained in the form of S_{11} vs f_r and Figure 8.8(b) represents that the antenna was resonating at 2.68 GHz whereas the corresponding return losses were −27.78 dB. The tested antenna resonates at a high frequency as compared to the simulated value i.e., 2.45 GHz which may be due to the effect of noise in the testing environment.

8.6 CONCLUSIONS

MFI results for the polymer composite represent that the MFI value will increase with the increase in wt. the proportion of PLA and decrease in CS and based upon that it may be concluded that the extrusion of the filament can be carried out at lower temperatures at a faster rate which was suitable for stretchable or flexible sensors fabricated via 3D printing.

- The thermal analysis of the composite material represents that the material was thermally stable while exhibiting less crystallization and hence suitable for the development of the implantable device.
- Concerning the mechanical properties of the composite, it was found that the material possesses great peak strength along with peak strain and is hence suitable for the fabrication of conformal sensors or antennas for biomedical applications.
- Based on morphological analysis, it was observed that there will be uniform mixing of the reinforced materials in the polymer, and the FTIR results highlight the presence of a strong double bond (which belongs to the carbonyl group) which may be responsible for the higher value of peak strength of the material.
- The actuation of the filament indicates that the prepared composite was responsive toward the hydro-based stimulus and hence may be utilized for the fabrication of smart sensors with additional shape memory properties.
- With regards to the RF characteristics, the sensor would resonate at a higher frequency when its surface was exposed to the SBF. From here it may be concluded that the losses would increase with the SBF and hence the sensor need to be redesigned to resonate in the ISM band.
- The RF behavior of the antenna indicates that it was suitable for the development of implantable purpose and hence resonates close to that of the simulated results, due to such capabilities, it may be utilized for health monitoring.

REFERENCES

Ahmed, B., Saleem, I., Zahra, H., Khurshid, H., Abbas, S. M. (2012). Analytical study on effects of substrate properties on the performance of microstrip patch antenna. *International Journal of Future Generation Communication and Networking*, 5(4):113–22.

Ariyapitipun, T., Mustapha, A., Clarke, A. D. (1999). Microbial shelf-life determination of vacuum-packaged fresh beef treated with polylactic acid, lactic acid, and nisin solutions. *Journal of Food Protection*, 62(8):913–20.

Banerjee, H., Suhail, M., Ren, H. (2018). Hydrogel actuators and sensors for biomedical soft robots: brief overview with impending challenges. *Biomimetics*, 3(3):15.

Burg, K. J., Porter, S., Kellam, J. F. (2000). Biomaterial developments for bone tissue engineering. *Biomaterials*, 21(23):2347–59.

Carver, K. and Mink, J. (1981). "Microstrip antenna technology", *IEEE Transactions on Antennas and Propagation* AP-29(1), 21.

Jayakumar, R., Prabaharan, M., Nair, S. V., Tamura, H. (2010). Novel chitin and chitosan nanofibers in biomedical applications. *Biotechnology Advances*, 28(1):142–50.

Kaur, G., Kaur, A., Toor, G. K., Dhaliwal, B. S., Pattnaik, S. S. (2015). Antennas for biomedical applications. *Biomedical Engineering Letters*, 5(3):203–12.

Kawakami, T., Antoh, M., Hasegawa, H., Yamagishi, T., Ito, M., Eda, S. (1992). Experimental
study on osteoconductive properties of a chitosan-bonded hydroxyapatite self-harden-
ing paste. *Biomaterials*, 13(11):759–63.

Kikuchi, M., Itoh, S., Ichinose, S., Shinomiya, K., Tanaka, J. (2001). Self-organization mech-
anism in a bone-like hydroxyapatite/collagen nanocomposite synthesized in vitro and
its biological reaction in vivo. *Biomaterials*, 22(13):1705–11.

Madihally, S. V., Matthew, H. W. (1999). Porous chitosan scaffolds for tissue engineering.
Biomaterials, 20(12):1133–42.

Menzel, W., and Grabherr W. (1991). "A microstrip patch antenna with coplanar feed line".
IEEE Microwave Guided Wave Lett., 1(11):340–2.

Nandiyanto, A. B., Oktiani, R., Ragadhita, R. (2019). How to read and interpret FTIR spectro-
scope of organic material. *Indonesian Journal of Science and Technology*, 4(1):97–118.

Paul, L. C., Hosain, M. S., Sarker, S., Prio, M. H., Morshed, M., Sarkar, A. K. (2015). The
effect of changing substrate material and thickness on the performance of inset feed
microstrip patch antenna. *American Journal of Networks and Communications*,
4(3):54–8.

Ranjan, N., Singh, R., Ahuja, I. P., Singh, J. (2019). Fabrication of PLA-HAp-CS based biocompatible and biodegradable feedstock filament using twin-screw extrusion. *Additive
manufacturing of emerging materials*, pp. 325–345.

Rasal, R. M., Hirt, D. E. (2009). "Micropatterning of covalently attached biotin on poly(lactic
acid) film surfaces", *Macro mol Bio sci*, pp. 989–96.

Ray, S. S., Okamoto, M. (2003). Biodegradable polylactide and its nanocomposites: opening a
new dimension for plastics and composites. *Macromolecular Rapid Communications*,
24(14):815–40.

Saleem, I. (2012). Analytical Evaluation of Tri-band Printed Antenna. *Information Sciences
Letters,* 1(3):3.

Singh, R., Kumar, R., Pawanpreet, Singh, M., Singh, J. (2022). On mechanical, thermal, and
morphological investigations of almond skin powder-reinforced polylactic acid feedstock filament. *Journal of Thermoplastic Composite Materials*, 35(2):230–48.

Singh, R., Kumar, S., Singh, A. P., Wei, Y. (2022). On comparison of recycled LDPE and
LDPE–bakelite composite based 3D-printed patch antenna. *Proceedings of the
Institution of Mechanical Engineers, Part L: Journal of Materials: Design and
Applications*, 236(4):842–56.

Smith, E. K. (1999). Radiowave propagation in ITU-R. *IEEE Antennas and Propagation
Magazine*, 41(1):118–9.

Sukhija, S., Sarin, R. K. (2017). Low-profile patch antennas for biomedical and wireless
applications. *Journal of Computational Electronics*, 16(2):354–68.

Ullah, S., Flint, J. A. (2014). Electro-textile-based wearable patch antenna on biodegradable polylactic acid (PLA) plastic substrate for 2.45 GHz, ISM band applications.
International Conference on Emerging Technologies (ICET), pp. 158–163.

Wan, A. C., Khor, E., Hastings, G. W. (1998). Preparation of a chitin–apatite composite by
in situ precipitation onto porous chitin scaffolds. *Journal of Biomedical Materials
Research: An Official Journal of The Society for Biomaterials, The Japanese Society
for Biomaterials, and the Australian Society for Biomaterials*, 41(4):541–8.

Zaidi, S. A., Tripathy, M. R. (2014). Design and Simulation-Based Study of Microstrip
E–Shaped Patch Antenna Using Different Substrate Materials. *Advance in Electronic
and Electric Engineering*, 4(6):611.

Index

Printed in the United States
by Baker & Taylor Publisher Services